UNDERSTANDING CIRCUITS

*Learning Problem Solving
Using Circuit Analysis*

Understanding Circuits, Learning Problem Solving Using Circuit Analysis
Khalid Sayood

ISBN: 978-3-031-00888-7 Sayood, Understanding Circuits (paperback)
ISBN: 978-3-031-02016-2 Sayood, Understanding Circuits (e-book)
Library of Congress Cataloging-in-Publication Data

First Edition
10 9 8 7 6 5 4 3 2 1

UNDERSTANDING CIRCUITS

Learning Problem Solving
Using Circuit Analysis

Khalid Sayood
Department of Electrical Engineering,
University of Nebraska,
Lincoln, USA

Abstract

This book/lecture is intended for a college freshman level class in problem solving, where the particular problems deal with electrical and electronic circuits. It can also be used in a junior/senior level class in high school to teach circuit analysis. The basic problem-solving paradigm used in this book is that of resolution of a problem into its component parts. The reader learns how to take circuits of varying levels of complexity using this paradigm. The problem-solving exercises also familiarize the reader with a number of different circuit components including resistors, capacitors, diodes, transistors, and operational amplifiers and their use in practical circuits. The reader should come away with both an understanding of how to approach complex problems and a "feel" for electrical and electronic circuits.

KEYWORDS

Circuit Analysis, Problem solving, Transistor Circuits, OpAmp Circuits, Diode Circuits

v

Contents

CHAPTER 1

Introduction

1.1   OVERVIEW

In this chapter we look briefly at our approach to problem solving.

1.2   PROBLEM SOLVING

Cutting up an ox

Prince Wen Hui's cook was cutting up an ox.

Out went a hand, down went a shoulder,

He planted a foot, he pressed with a knee,

The ox fell apart with a whisper,

The bright cleaver murmured Like a gentle wind.

Rhythm! Timing! Like a sacred dance,

Like "The Mulberry Grove," Like the ancient harmonies!

"Good work!" the Prince exclaimed, "Your method is faultless!"

"Method?" said the cook laying aside his cleaver,

What I follow is Tao, beyond all methods!

"When I first began to cut up oxen

I would see before me the whole ox

All in one mass, "After three years

I no longer saw this mass I saw the distinctions.

"But now, I see nothing with the eye. My whole being
Apprehends. My senses are idle. The spirit
Free to work without plan follows its own instinct
Guided by natural line, By the secret opening, the hidden space,
My cleaver finds its own way. I cut through no joint, chop no bone.

"A good cook needs a new chopper Once a year - he cuts.
A poor cook needs a new one Every month - he hacks!

"I have used this same cleaver nineteen years.
It has cut up a thousand oxen.
Its edge is as keen as if newly sharpened

"There are spaces in the joints; The blade is thin and keen:
When this thinness finds the space
There is all the room you need! It goes like a breeze!
Hence I have this cleaver nineteen years as if newly sharpened!

"True, there are sometimes tough joints. I feel them coming,
I slow down, I watch closely, hold back, barely move the blade,
And whump! the part falls away landing like a clod of earth.

Then I withdraw the blade, I stand still
And let the joy of the work sink in.
I clean the blade and put it away."

Prince Wan Hui said, "This is it! My cook has shown me
How I ought to live My own life!"

From *The way of Chuang Tzu* by Fr. Thomas Merton

This poem might seem like a strange way to start a book on problem solving in electrical engineering. It is not, after all, a course on oriental philosophy or the musings of monks; nor, for that matter, is it a course on slaughtering livestock. However, this is a book on problem solving, and this poem is as good an analogy as any that I have found to illustrate how you go about solving problems in engineering. Any problem deserving of its designation initially looks like a rather large, overbearing, and clearly unresolvable mass. It is generally understood that the problem has to be broken down into smaller pieces before a solution can be found. But how do you go about breaking a problem into smaller pieces? You can hack away at the problem until, finally, dripping with sweat and gore, you succeed in breaking it down into pieces small enough. There are several problems with this approach. It takes a long time, it tires you out so you make mistakes, and it just does not feel cool. And what, you say, has feeling cool got to do with solving problems? Well, look at it this way. Your intent at this point in your life is to become an engineer—a profession that you probably plan to follow for a major part of the rest of your life. Being an engineer means solving problems, and if solving problems gives you a cool feeling you will have a lot of fun. If not, you are faced with a lifetime of drudgery.

We want an approach to problem solving that reduces the amount of hacking required. We prefer an approach that, in the words of the poem, allows us to see the spaces between the pieces of the problems, and permits us to use the tools at our disposal to separate the massive thing into understandable pieces. Breaking a problem into smaller pieces is such a standard approach to solving problems that it has a name. It is called *analysis*.

As the problems we are trying to resolve are not oxen, what we mean by the spaces between the pieces, and what we mean by tools are very different from what Prince Wen Hui's cook meant by them. The spaces between the problems are made clear by the rules that govern physical systems. Each part of a system interacts with other parts of the system according to a set of rules. By discovering and using this set of rules we can see where our points of attack should be. Our tools are whatever we use to separate, hold apart, and recombine the pieces of the problem.

In these notes we will try to understand the process of analysis by using circuits. The reason for this is twofold. One reason is that we want to introduce you to some aspects of electrical engineering. The second is that the rules by which different parts of an electrical circuit interact are very simple and it is relatively easy to see the spaces. As we will see in the next chapter, there are only two laws that govern the behavior of electric circuits. Together with the rules that govern individual components, these two laws are enough to permit us to analyze circuits. As you learn to break down a problem into its component pieces you will find that even the most intimidating circuit can be attacked and resolved in this manner.

Electrical engineering is a very diverse field. Look around you. Products of electrical engineers are ubiquitous. The particular problems that we look at in these notes belong to one small part of electrical engineering. Depending on the branch of electrical engineering you later pursue, you may or may not use the information provided in this book. However, I hope to give you a vision and a feel for engineering which will be helpful to you regardless of your particular specialization.

C H A P T E R 2

Current, Voltage, and Circuit Laws

2.1 OVERVIEW

In this chapter we introduce the concepts of current and voltage and present two laws that govern all electrical circuits.

2.2 ELECTRICITY, CURRENT, AND VOLTAGE

Living in the modern world means you have considerable familiarity with the uses of electricity. You may also have received a formal introduction in your physics classes. Electricity is (literally) in the air. Consider a simple crystal radio circuit consisting of a coil of wire wound around an empty film cannister, an earphone, and a diode (see Fig. 2.1).

Notice that there are no batteries in this circuit. Yet, by connecting the ground wire to a cold water pipe (or the cable outlet), and the antenna wire to a long wire or to yourself, you should be able to hear one or more of the nearby AM radio stations. This is because the signal from radio stations, and TV stations, and cell phones, are in the air all around you. The circuit converts these signals into a form you can hear. Try this! A detailed parts list is as follows:

- 30 gauge magnet wire for tuning coil;

- one plastic film can for winding the coil;

- one germanium diode;

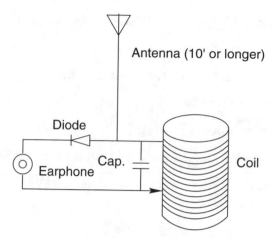

FIGURE 2.1: A crystal radio.

- one high impedance earphone;

- one 470 pF capacitor;

- 15 ft of wire for antenna and ground connections.

If you do not want to go around looking for the parts, crystal radio kits are available in most electronics stores.

Understanding the basic nature of electricity is beyond the scope of this book. Rather, our goal is to get enough of an understanding of the behavior of electricity in order to understand how electrical circuits function. We describe the behavior of electricity using the concepts of current and voltage.

Current is a measure of the amount of charge flowing past a given point per second and is measured in units of amperes (A). One ampere of current is the flow of one coulomb of charge per second. An analogy to this is the flow of water past a given point. The greater the flow, the higher the current. We often deal with currents that are on the order of a thousandth of an ampere or a millionth of an ampere. Using the metric naming conventions, we call a thousandth of an ampere a milliampere (mA) and a millionth of an ampere a microampere (μA).

Voltage is the difference in potential energy between two points. Formally, the voltage between two points is the amount of work done in moving one coulomb of charge from the lower potential point to the higher potential point.

Use of the words "higher" and "lower" suggest an analogy for voltage. In physics you may have learned that energy is the ability to do work. If we raise a rock to a height of h_1 meters above the floor, the rock has some energy stored in it. This energy can be used by releasing the rock. When the rock is released it will hit the floor with a force proportional to its original height above the floor. Thus the energy stored in the rock (called potential energy) is directly proportional to its height above the floor. If we denote the potential energy by E_p, then

$$E_p = mgh_1$$

where m is the mass of the rock and g is the acceleration due to gravity. Now suppose we place a table of height h_2 meters between the rock and the floor. If we release the rock it will travel a distance of $h_1 - h_2$ meters before it hits the table. As it travels a smaller distance ($h_1 - h_2$ meters instead of h_1 meters) the force with which it hits the table will be less than the force with which it would have hit the floor. This is because its potential energy with respect to the table was $mg(h_1 - h_2)$, which is less than mgh_1. If we examine the expression of the potential energy of the rock, we can write it as a difference of two potential energies:

$$mg(h_1 - h_2) = mgh_1 - mgh_2$$

i.e., the potential energy of the rock at height h_1 and the potential energy of the rock at height h_2. It is the difference in potential that allows the rock to do work.

The potential energy of the rock arises because of the work done against gravity in raising it. The potential energy of interest to us in electrical circuits is a result of work done against an electric field. The voltage difference, or voltage between two points, is the difference in this electrical potential between two points.

This difference in potential can be used to make current flow between two points. In terms of our water analogy, think of voltage as the difference in potential energy of water at different heights. Water will not flow between containers at the same level (see Fig. 2.2); however, it will flow when the containers are at different heights (see Fig. 2.3).

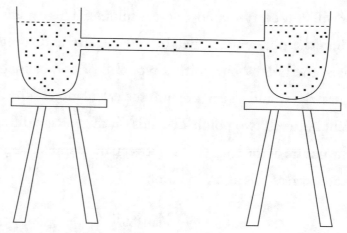

FIGURE 2.2: Nonflowing water.

Similarly, we can have two points that have a high electric potential with respect to a reference or ground point, but whether current will flow between these points is a function of two things: whether there is a path for the current to flow and whether there is a difference in potential.

Consider birds sitting on a high-voltage line. There is certainly a path for the current to flow through, but the potential difference between the two points at which the birds' feet are touching the wire is so small as to be negligible. Consequently, no current flows through them, and the birds do not get cooked.

If we have two containers of water at different heights and we connect them with a pipe, water will flow from the higher container to the lower container. But

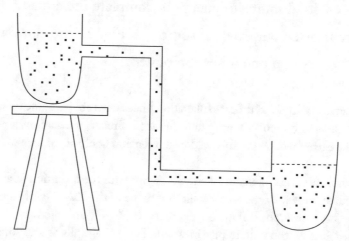

FIGURE 2.3: Flowing water.

how much water will flow every second, or every minute? This depends on a number of things including the difference in height and the size and composition of the pipe. If we replace the hollow pipe with a pipe full of porous material, then the porosity of the material would also figure into the calculation of the flow. Similarly, the flow of current between two points depends on the voltage difference between the points and the nature of the connection between the points. Using mathematical shorthand, we can express this as an equation:

$$I_{ab} = f(N_{ab}, V_{ab})$$

where V_{ab} denotes the voltage difference between points a and b, I_{ab}[1] denotes the current from point a to point b, N_{ab} denotes the nature of the connection between points a and b, and $f(\ )$ denotes the functional relationships. We will explore the functional relationship between the voltage across a component and the current through it for a number of different components in the following chapters. In this chapter, we will concentrate on developing universal rules or laws that hold for all components in an electrical circuit.

The word circuit comes from the Latin *circumire*, which means to go around. In our analogy, water flows from one point to another. For the water to go around we add a little rotating gadget, as shown in Fig. 2.4, which transfers water from the lower bucket to the upper bucket, thus completing the *circuit*. Obviously there is a need for an external source of energy to complete the circuit.[2] In an electrical circuit, a battery is often our external source. In Fig. 2.5 we show a simple circuit consisting of a battery and one other component.

---

[1] Why represent current with I? The best explanation I have been able to come up with is that current used to be referred to as current intensity and, in fact, a French translation of current is *intensite*. Since many of the earlier workers in this field were French, I could stand for the *intensity* in current intensity.

[2] In a sense this whole business of electrical circuits begins with the invention of an external (more or less) constant source known by the rather inelegant name of the *Voltaic pile* invented by Allesandro Volta. Volta was born in Como, Italy, on February 18, 1745, and died on March 5, 1827. He published his account of the voltaic pile in 1800. The voltaic pile is the forerunner of today's batteries (in French as well as several other languages, a battery is called a *pil*).

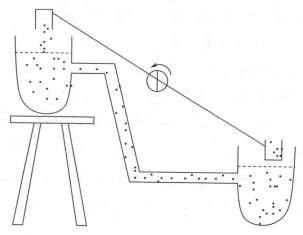

FIGURE 2.4: A water circuit.

At this point let's pause to note a few conventions. When we refer to a difference between two values, our knowledge of the situation is incomplete unless we know which of the two values is larger. In an electrical circuit, the size of the difference in potential (along with the nature of the component) determines the size of the current. However, the direction of the current through a component is dictated by which end of the component is at a higher potential. Thus, it is important not only to know the size of the voltage difference across a component but also which terminal is at a higher potential level.

Recall our water analogy: if we say the difference in water level between pail 1 and pail 2 is 2 ft, we cannot tell whether water will flow from pail 1 to pail 2 or vice versa. However, if we know which one is higher, then we can tell which direction the water will flow.

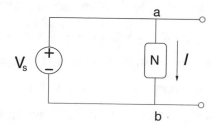

FIGURE 2.5: An electrical circuit.

If pail 1 is at a height of 3 ft and pail 2 is at a height of 5 ft, we can say that *pail 2 is higher than pail 1 by 2 ft*. If their places are reversed, we can say *pail 1 is higher than pail 2 by 2 ft*. We can also do the following: In the first case with pail 1 at 3 ft and pail 2 at 5 ft, we say *pail 1 is higher than pail 2 by −2 ft*, and in the second case we say *pail 1 is higher than pail 2 by 2 ft*. We have fixed the place of pail 1 and pail 2 in our sentence so that our sentence will always read *pail 1 is higher than pail 2 by x feet*, where x can be positive or negative. This might seem rather cumbersome but, as you will see, this brings a precision to our statements, which is very useful in solving circuits.

In an electrical circuit, for a given component we denote the terminal that we assign to be at a higher potential with a positive (+) sign. The other terminal is marked with a negative (−) sign. When we say that the voltage across a component is V, we mean that the difference between the potential at the terminal marked + (the positive terminal) and the terminal marked − (the negative terminal) is V volts.

The component N and the voltage source in Fig. 2.5 are connected by wires. We assume that the potential along the wires is always the same. That is, the potential difference between any two points on the wire is 0. Therefore, in the circuit shown in Fig. 2.5, the voltage across the component N is simply V_s. If we knew the relationship between the voltage and the current for the component N, we could have used that to obtain the current I through the circuit. Notice in the figure that we have also assigned a direction to the current. This is imperative as *we cannot talk about voltages and currents without assigning polarities and directions, respectively.*

Suppose V_s in Fig. 2.5 is 9 V. This means that the potential difference between point a and point b is 9 V, with point a being at the higher potential. We write this as $V_{ab} = 9$ V. Suppose we were asked for V_{ba}. This is the difference in potential between point b and point a. As point a is at a higher potential $V_{ba} = -9$ V. Reversing the order of the subscript negates the value.

Similarly, suppose we had a current of 2 A flowing in the circuit from a to b. In other words, $I_{ab} = 2$ A. However, suppose we were interested in the current flow

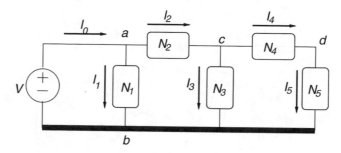

FIGURE 2.6: Example of an electrical circuit.

from b to a. In this case, we would obtain a current value of -2 A, or $I_{ba} = -2$ A. Notice that the subscript for the current denotes the direction of current flow; I_{ab} is the current from a to b and I_{ba} is the current from b to a. As in the case of the voltage, reversing the order of the subscript results in negating the value of the current.

2.3 KIRCHHOFF'S LAWS

When we have a single source and single connecting component, it is a simple matter to obtain the current through the circuit and the voltages between various points. If we have more components and/or sources such as in Figs. 2.6 or 2.7 we need a few more rules. These rules were developed by Gustav Kirchhoff[3] and are known as Kirchhoff's Laws or more specifically Kirchhoff's current law and Kirchhoff's voltage law. These laws govern how all electrical circuits behave. Although they are very simple, they are essential tools for analyzing circuits. The Kirchhoff's laws can easily be understood in terms of our water analogy.

2.3.1 Kirchhoff's Current Law

Kirchhoff's current law (KCL) relates to currents entering and leaving a *node*. A node is a junction in the circuit where two or more components are connected. The word "junction" is used somewhat loosely here. Consider Fig. 2.6. The voltage source and the components N_1 and N_2 are connected at node a. The components

---

[3] Gustav Robert Kirchhoff was born in Koenigsberg, Germany, on March 12, 1824, and died on October 17, 1887. Working with Bunsen (of the Bunsen burner fame), Kirchhoff also discovered that glowing vapors absorbed light of definite wavelengths.

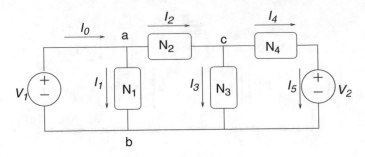

FIGURE 2.7: Another example of an electrical circuit.

N_2, N_3, and N_4 are connected at node c. In both of these cases, the nodes correspond to points on the circuit diagram. However, node b at which the components N_1, N_3, N_5, and the source come together is the bold line shown at the bottom of the figure.

Kirchhoff's law states that the current entering a node is always equal to the current leaving a node. If we consider the analogous situation in our water example, we can see that this has to be true. Consider the situation shown in Fig. 2.8. If there is a continuous flow of water, obviously the amount of water entering the junction is equal to the amount of water leaving the junction.

In the part of a circuit shown in Fig. 2.9, the current entering the node b is the current I_{ab}. The currents leaving node b are the currents I_{bc} and I_{bd}. Therefore, according to the KCL

$$I_{ab} = I_{bc} + I_{bd}$$

A simple way of checking whether you have written an equation correctly is to check the subscripts. Given our convention the second subscript of all currents

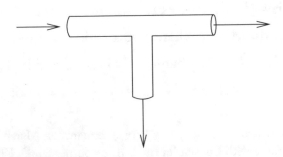

FIGURE 2.8: A joining of pipes.

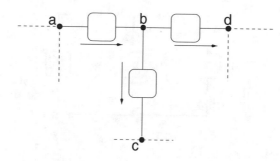

FIGURE 2.9: A part of a circuit.

entering a node x should be x, while the first subscript of all currents leaving node x should be x.

Kirchhoff's current law can also be stated as *the algebraic sum of all currents entering a node is zero.* If we look at node c in the circuit shown in Fig. 2.6, the KCL says that

$$I_2 = I_3 + I_4 \qquad (2.1)$$

In terms of our earlier naming convention, this is the same as saying

$$I_{ac} = I_{cb} + I_{cd}$$

The currents entering node c are I_{ac}, I_{bc}, and I_{dc}. Therefore, according to the second statement of the KCL

$$I_{ac} + I_{bc} + I_{dc} = 0 \qquad (2.2)$$

At the end of the previous section we said that reversing the order of the subscripts results in negating the current value. Therefore, if $I_{cb} = I_3$, then $I_{bc} = -I_3$, and if $I_{cd} = I_4$, then $I_{dc} = -I_4$. Substituting these in Eq. (2.2) we get

$$I_2 - I_3 - I_4 = 0 \qquad (2.3)$$

which is the same as Eq. (2.1). Both statements of the KCL mean exactly the same thing.

Let's see how we would use the KCL in practice.

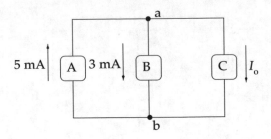

FIGURE 2.10: An application of KCL.

Example 2.3.1: Consider the circuit shown in Fig. 2.10. There are three components A, B, and C connected to form a circuit. Suppose we measure the current through A and B and come up with the measured values indicated in the figure. What is the value of the current marked I_o?

Following the example of Prince Wen Hui's cook, the first thing we do is look for the spaces between the components of the problem. We know the behavior of currents at nodes; therefore, in the case of electrical circuits, the spaces are generally the nodes. In this circuit we have only two nodes named a and b. Our "knife" is the current law and we can apply it to either of the nodes. Let's apply it to the node labeled b. At this node, the current with a value of 3 mA flows *to* node b through the component marked B, and 5 mA flows *from* node b through the component marked A. Because the current marked I_o is the only other current entering the node, it has to have a value that will make the total current entering the node equal to the total current leaving the node. That is,

$$I_o + 3\,\text{mA} = 5\,\text{mA}$$

or I_o is 2 mA.

Let's look at one more circuit which, at first glance, looks somewhat more complicated.

Example 2.3.2: Consider the circuit shown in Fig. 2.11. We have five components instead of three and twice the number of nodes we had in the previous example. However, it is actually no more difficult to solve than the previous problem. The

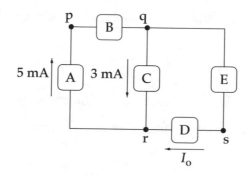

FIGURE 2.11: Another application of KCL.

important thing is not to get distracted by the size of the circuit and concentrate on the nodes. In this problem the unknown current I_o enters node r; therefore, we look at the currents entering or leaving this node. From the figure, it is clear that the currents entering node r are I_o and 3 mA, and the current leaving node r is 5 mA. Therefore, once again I_o has a value of 2 mA.

Finally, let's look at a circuit where we need multiple applications of the KCL to get our answer.

Example 2.3.3: Consider the circuit shown in Fig. 2.12. This circuit has considerably more components than the previous two. Again, we wish to find the current marked I_o. This is the current leaving node f and entering node g. We can apply the current law to either of these nodes to begin the process of finding the value of I_o. Let's apply KCL to node f. The current entering node f is I_{df} and the current

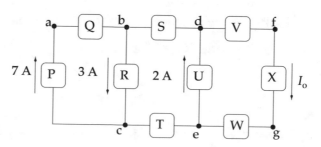

FIGURE 2.12: Another application of KCL.

leaving node f is I_{fg}, which is the same as I_o. Therefore,

$$I_o = I_{fg} = I_{df}$$

To find I_{df} we apply KCL to node d. The currents entering node d are I_{bd} and I_{ed}, while the current leaving node d is I_{df}. Therefore,

$$I_{df} = I_{bd} + I_{ed}$$

From the figure we can see that $I_{ed} = 2\,\text{A}$. Therefore,

$$I_o = I_{df} = I_{bd} + 2 \tag{2.4}$$

To find I_{bd} we apply KCL to node b. The current entering node is I_{ab}, while the currents leaving node b are I_{bc} and I_{bd}. Therefore,

$$I_{ab} = I_{bc} + I_{bd}$$

or

$$I_{bd} = I_{ab} - I_{bc}$$

As $I_{bc} = 3\,\text{A}$

$$I_{bd} = I_{ab} - 3$$

Substituting this expression for I_{bd} in Eq. (2.4) we obtain

$$I_o = I_{ab} - 3 + 2 = I_{ab} - 1 \tag{2.5}$$

To find I_{ab} we look at node a, where I_{ab} is the current leaving the node and I_{ca} is the current entering the node. Therefore,

$$I_{ab} = I_{ca}$$

From the figure $I_{ca} = 7$, therefore, $I_{ab} = 7$. Substituting this value for I_{ab} into Eq. (2.5) we get $I_o = 6$.

Quite a few steps were necessary to obtain the solution in this last example. However, notice that each step of the solution was a very straightforward application of a simple rule. As we study more and more complicated *looking* problems, it will be extremely important that we do not get distracted by the complexity of the overall problem and instead focus on the simplicity of the steps needed to solve the problem. Simple step by simple step, we will be able to solve the most difficult problems.

2.3.2   Kirchhoff's Voltage Law

Kirchhoff's voltage law (KVL) states that the voltage difference between two points is the same regardless of the path we take between them. Let's make use of our height analogy to get a mental picture of what we mean by the KVL. Consider the situation shown in Fig. 2.13. The height difference between level a and level c is h_1. If we place a rock of mass m at level a, the potential energy of the rock with respect to level c will be mgh_1. If we go from level a to level b the change in height is h_2. The change in height from level b to level c is h_3. Clearly

$$h_1 = h_2 + h_3$$

and

$$mgh_1 = mgh_2 + mgh_3$$

Whether we go directly from a to c, or go first to b and then to c, the change, or difference, in potential energy is the same.

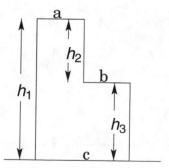

FIGURE 2.13: A case of heights.

In Fig. 2.6 the voltage difference between nodes a and c is denoted by V_{ac}. We could go from node a directly to node c via component N_1, or we can go from node a to node b and from there to node c. The voltage difference between nodes a and b is V_{ab}, and the voltage difference between nodes b and c is V_{bc}. According to the KVL,

$$V_{ab} = V_{ac} + V_{cb}$$

A consequence of this is that if we traverse any closed path in a circuit, the algebraic sum of the voltages is zero. Consider the circuit in Fig. 2.6 and the path acba. V_{aa} is the potential difference between point a and point a, which is zero (just as the difference in height between level a and level a is zero). We can get from a to a by going first to b then to c and then back to a. Therefore,

$$V_{aa} = V_{ab} + V_{bc} + V_{ca} = 0$$

Example 2.3.4: Consider the circuit in Fig. 2.14. The voltage difference between nodes a and c is 9 V, while the voltage difference between nodes b and c is 6 V. In other words,

$$V_{ac} = 9\,\text{V}$$
$$V_{bc} = 6\,\text{V}$$

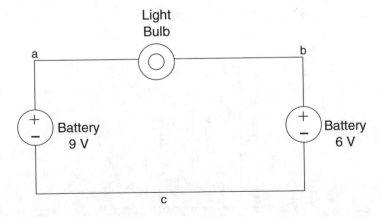

FIGURE 2.14: Two batteries and a lamp.

We wish to find the voltage across the lamp. By KVL

$$V_{ac} = V_{ab} + V_{bc}$$

or

$$V_{ab} = V_{ac} - V_{bc}$$

which means that $V_{ab} = 3$ V.

We can go from b to c in two different ways, directly or via node a. Therefore, we can write

$$V_{bc} = V_{ba} + V_{ac}$$

from which we obtain $V_{ba} = -3$ V. As with the case of the currents, notice that reversing the subscript results in a negation of the value and $V_{ab} = -V_{ba}$.

Let's continue with another example, which is slightly more complicated.

Example 2.3.5: Consider the circuit shown in Fig. 2.15 consisting of four components labeled A, B, C, and D. The voltage across component A is 5 V and across component D is 12 V. We are asked to find the voltage labeled V_o across the component labeled B. We first identify the nodes in this circuit and label them x, y, and z. (Can you see why there are only three nodes in this circuit?) The voltage across the component labeled A is the potential difference between nodes z and x with node z being at the higher potential. That is, $V_{zx} = 5$ V. Similarly we can see that

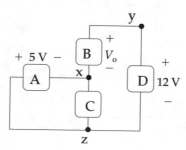

FIGURE 2.15: An application of KVL.

$V_{yz} = 12$ V and the unknown voltage V_o is the same as V_{yx}. Using KVL,

$$V_{yx} = V_{yz} + V_{zx} = 12 + 5 = 17 \text{ V}$$

Although initially the problem looked complicated, it really was not.

Finally, let us take an example from everyday life.

Example 2.3.6: Suppose your car battery runs down. Instead of the required 12 V across it, you are getting 10 V, so you ask for a jump from a friend. The correct procedure is to connect the positive terminals of the batteries (the negative terminals are already connected to the car chassis), and connect the chassis(s) to each other. As you do so, right before you connect the chassis, the situation is similar to as shown in Fig. 2.16. We want to find the voltage V_g across the gap. At the bottom of

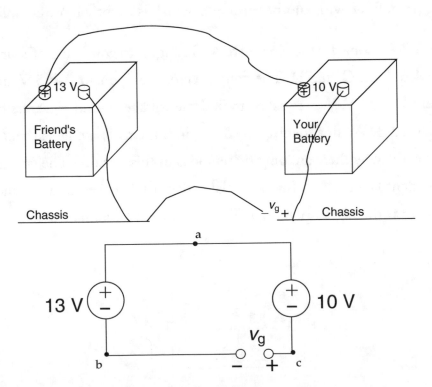

FIGURE 2.16: Correct way to connect the batteries.

Fig. 2.16 we have redrawn the circuit and identified and labeled the nodes. From this figure we can see that V_g is the same as V_{cb}, the potential difference between nodes c and b. We know the voltages V_{ab} (13 V) and V_{ac} (10 V). In order to find V_{cb} we need to write this voltage in terms of the known voltages. Using the KVL, we can write V_{cb} as

$$V_{cb} = V_{ca} + V_{ab}$$

We know that V_{ac} is 10 V; therefore, V_{ca} is -10 V and

$$V_{cb} = -10 + 13 = 3 \text{ V}$$

Thus, the voltage across the gap is 3 V.

What happens when we incorrectly connect the negative terminal of your friend's battery to the positive terminal of your battery as shown in Fig. 2.17? Again we have redrawn the circuit and identified and labeled the nodes in the bottom half of the figure. The voltage across the gap V_g is still V_{bc}, and V_{ac} is still 10 V. However,

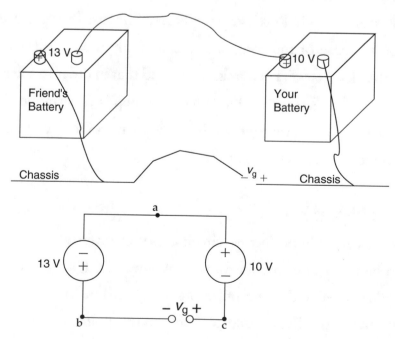

FIGURE 2.17: Incorrect way to connect batteries.

in this configuration V_{ab} is -13 V. As V_{cb} is still given by

$$V_{cb} = V_{ca} + V_{ab}$$

the voltage across the gap is

$$V_g = V_{cb} = -10 - 13 = -23\,\text{V}$$

This is not a safe situation.

In the examples provided for the KCL, you may have noticed that we had circuits in which some of the currents were known and, in the examples provided for the voltage law, the voltages were known. In practice (for a given circuit) we may know voltages across some components and currents through others. In order to apply the current law or the voltage law, we need to convert the voltages across the components to the currents through them, or vice versa. For this to happen we need to know the relationship between the voltage across a particular component and the current through it. We call this relationship the *component rule*. In the following chapters, we will see how we can use these component rules and the Kirchhoff's laws to solve what initially look like complicated circuits. In the process we will introduce you to some of the components commonly used in electrical circuits. Throughout the analysis you need to keep in mind that in order to solve a problem you may have to take many steps. However, each step is very simple. Like Prince Wen Hui's cook, if you concentrate on the spaces, which for us will mainly be the nodes, the problem will resolve itself simply.

2.4   SUMMARY

In this chapter we introduced the two circuit laws that govern all circuits and will be with us from here on. The Kirchhoff's current law or KCL states that the current entering a node is equal to the current leaving the node. The Kirchhoff's voltage law or KVL states that the voltage difference between two points is independent of the path taken to get from one point to the other. These are very simple statements and it would be easy to underestimate their power. We will find in later chapters that these two laws along with rules describing the behavior of individual components is all we need to analyze even the most complex circuit.

2.5   PROJECTS AND PROBLEMS

1. In the circuit shown below find I_o.

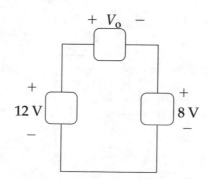

2. In the circuit shown below find I_o.

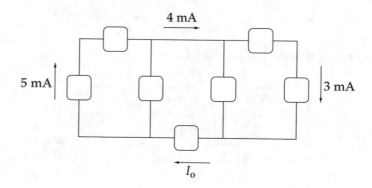

3. In the circuit shown below find V_o.

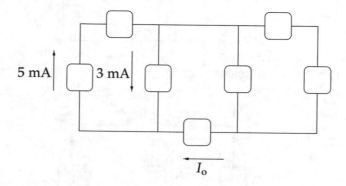

4. In the circuit shown below find V_o.

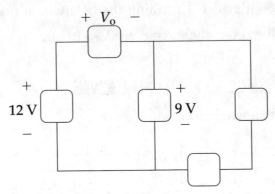

5. In the circuit shown below find I_o.

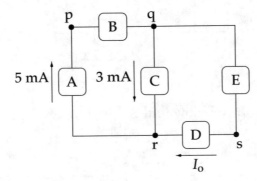

6. In the circuit shown below find I_o.

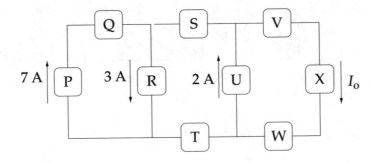

7. In the circuit shown below find V_o.

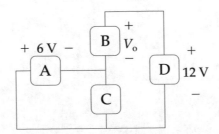

8.   In the circuit shown below find V_o.

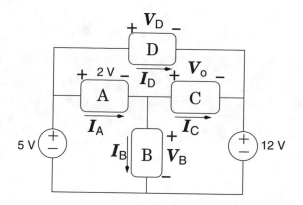

9.   In the circuit shown below find V_o, V_1, and V_2.

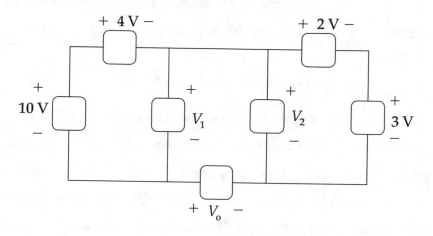

C H A P T E R 3

Resistive Circuits

3.1   OVERVIEW

In this chapter we see how we can use the Kirchhoff's laws together with component rules to analyze circuits. We also begin looking at various components used in electrical circuits. We start with the simplest component—the resistor.

3.2   INTRODUCTION

In the previous chapter we described the two laws that we will use to analyze all kinds of circuits. One law deals with currents at a node and the other with voltage between two points. Also, we ignored the nature of the components and assumed that we somehow know the currents and voltages necessary to obtain the information we desire. In order to apply the Kirchhoff's laws we often need to know the relationship between the current through the component and the voltage across it. As the circuits become more complicated it is also useful to have a procedure for attacking the problem. Like Prince Wen Hui's cook, we have to identify the spaces and then separate the problem along these spaces. In circuits the spaces will be the nodes in the circuit. The tools we use at these spaces will be the Kirchhoff's laws along with the current–voltage relationship for the component. Let's develop our procedure using a simple example.

Example 3.2.1: Consider the circuit shown in Fig. 3.1, which consists of the mystery components A, B, and C. We do not know what these components are. All we know

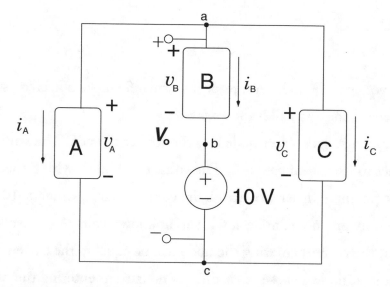

FIGURE 3.1: A collection of boxes.

is that they have the following current–voltage relationships:

$$I_A = 0.1\, V_A \tag{3.1}$$

$$I_B = 0.05\, V_B \tag{3.2}$$

$$V_C = 2.5 + 10\, I_C \tag{3.3}$$

Is this sufficient to compute V_o? Let's find out.

Let's start with counting the nodes. There are three nodes in this circuit. One is at the top of the figure where components A, B, and C come together. One is between component B and the 10-V source, and one is at the bottom of the figure where components A, C, and the voltage source are joined. Let's name these nodes a, b, and c. We will attempt to find the voltage at each node with respect to a reference node. Once we know the voltage of all nodes with respect to a reference node the voltage difference between any two nodes can always be found as the difference between the voltages of those nodes with respect to the common node. For example, suppose we know V_{bx} and V_{ax}, where x is the common node. Suppose we wished to find V_{ab}. Using KVL we know that

$$V_{ax} = V_{ab} + V_{bx}$$

or

$$V_{ab} = V_{ax} - V_{bx}$$

Because the voltage of a node with respect to the reference node is so useful we will give it a special name and call it the *node voltage*.

In our circuit, let's select node c as the reference node. The voltage of node c with respect to itself is of course 0. The voltage of node b with respect to node c is simply the potential difference across the voltage source, which is 10 V. And we are left with only one node, node a, with an unknown voltage. All we know about a node is that the current entering the node will be equal to the current leaving the node. Looking at node a we see that there is no current entering the node and we have I_A, I_B, and I_C leaving the node. Thus,

$$I_A + I_B + I_C = 0$$

We would like to write this equation in terms of the voltages at each node with respect to the reference node. We will get there in two steps. First, write the currents through the components in terms of the voltages across the components. We can do this by using the component rules specified in Eqs. (3.1)–(3.3). We replace I_A with $0.1\,V_A$, I_B with $0.05\,V_B$ and I_C with $(V_C - 2.5)/10$ to obtain

$$0.1\,V_A + 0.05\,V_B + \frac{V_C - 2.5}{10} = 0 \tag{3.4}$$

From the circuit we can see that

$$V_A = V_{ac}$$
$$V_B = V_{ab}$$
$$V_C = V_{ac}$$

Substituting these into Eq. (3.4) we obtain

$$0.1\,V_{ac} + 0.05\,V_{ab} + \frac{V_{ac} - 2.5}{10} = 0 \tag{3.5}$$

In this equation V_{ac} is the voltage of node a with respect to the reference node; however, V_{ab} is not a voltage with respect to the reference node. As noted earlier we can write the voltage between any two nodes in terms of the voltages of those nodes with respect to a common node. Thus,

$$V_{ab} = V_{ac} - V_{bc}$$

Noting that $V_{bc} = 10$, we obtain

$$V_{ab} = V_{ac} - 10$$

Substituting this in Eq. (3.5) we get

$$0.1\,V_{ac} + 0.05(V_{ac} - 10) + \frac{V_{ac} - 2.5}{10} = 0 \qquad (3.6)$$

which is all in terms of V_{ac}. Solving for V_{ac} we obtain $V_{ac} = 3$ V.

We had originally wanted to compute V_o. Looking at Fig. 3.1 we can see that $V_o = V_{ac}$. Therefore $V_o = 3$ V.

In this example we have applied the Kirchhoff's laws in a systematic way. Systematic procedures are nice to know because they can be used to solve a large number of problems. They allow us to concentrate on the spaces rather than on the whole problem. Let's write down the procedure used in the previous example in general terms. The steps we took were as follows:

1. Identify all nodes in the circuit.

2. Select one of these nodes as the reference node.

3. At each node, which is not the reference node, either
   (a) write the node voltage, the potential difference between the node and the reference node, **or,**
   (b) write the Kirchhoff's current law at the node.

4. Wherever possible, write the currents in the equations in terms of the voltages across the components.

5. Write the voltages in terms of the node voltages, the voltage of each node with respect to the reference node.

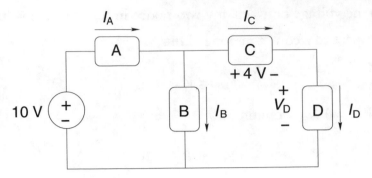

FIGURE 3.2: Another collection of boxes.

If you have a total of N nodes, at this point you will have at most $N-1$ equations with $N-1$ unknowns. The unknowns will be the potential difference between all nodes and the reference node. You can solve these equations with whatever tools you have at your disposal.

This procedure will work for most of the circuits you encounter. For the remaining, we will have to modify the procedure. Let's look at an example where we do have to modify the procedure.

Example 3.2.2: Consider the circuit shown in Fig. 3.2. Suppose we are given the following rules for the components:

$$I_A = V_A \tag{3.7}$$

$$I_B = 0.125\, V_B \tag{3.8}$$

$$I_D = 0.25\, V_D \tag{3.9}$$

and asked to find the voltage across component D. Let's follow the steps of our procedure as far as we can.

The first step is to identify the nodes. We have done so in the Fig. 3.3 and marked them with the letters a, b, c, and d. We have also relabeled the currents. The current I_A is the current going from node a to node b, so we have labeled it I_{ab}. Similarly the current I_B is the current from node b to node d and is therefore labeled I_{bd}. The same is true for I_C and I_{bc}, and for I_D and I_{cd}. Next we have to pick a reference node. Let's pick that to be node d. Now let's visit each node.

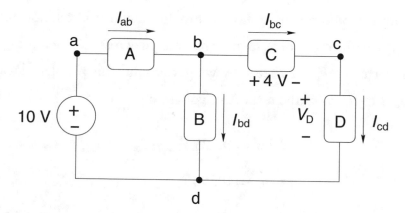

FIGURE 3.3: Labeled collection of boxes.

node a: We can write the voltage at node a with respect to node d by inspection.

$$V_{ad} = 10\,\text{V}$$

node b: At node b we can write the current law as

$$I_{ab} = I_{bd} + I_{bc} \tag{3.10}$$

node c: At node c the current entering the node is I_{bc} and the current leaving the node is I_{cd}. Therefore we can write the current law as

$$I_{bc} = I_{cd} \tag{3.11}$$

We have two equations in terms of the currents. Now, let's try and write these currents in terms of the voltages across the components. The current I_{ab} is the current through component A. From Eq. (3.7) we can see that the current through component A is equal to the voltage across the component

$$I_{ab} = V_{ab}$$

The current I_{bd} is the current through component B. From Eq. (3.8)

$$I_{bd} = 0.125\,V_{bd}$$

Similarly, using Eq. (3.9)

$$I_{cd} = 0.25\,V_{cd}$$

However, we do not have any information about the relationship between the current through component C and the voltage across it. Therefore, we cannot write I_{bc} in terms of V_{bc}. To see how we get around this problem let's write the other currents in terms of the respective voltages:

$$V_{ab} = 0.125\, V_{bd} + I_{bc} \tag{3.12}$$

$$I_{bc} = 0.25\, V_{cd} \tag{3.13}$$

We can see that we can substitute the expression for I_{bc} from Eq. (3.13) in Eq. (3.12) to obtain

$$V_{ab} = 0.125\, V_{bd} + 0.25\, V_{cd} \tag{3.14}$$

Let's now rewrite this equation in terms of the voltages at each node with respect to the reference node.

$$V_{ad} - V_{bd} = 0.125\, V_{bd} + 0.25\, V_{cd} \tag{3.15}$$

Substituting the value for V_{ad} we get

$$10 - V_{bd} = 0.125\, V_{bd} + 0.25\, V_{cd} \tag{3.16}$$

We have two unknowns, V_{bd} and V_{cd}, and only one equation! However, we have not used all the information available to us. Looking at Fig. 3.2 we see that the voltage across component C is 4 V, or $V_{bc} = 4$. Writing V_{bc} in terms of voltages in terms of the reference node we get

$$V_{bd} - V_{cd} = 4\,\text{V} \tag{3.17}$$

which is our second equation. Given two equations with two unknowns we can solve for V_{bd} and V_{cd} and obtain

$$V_{bd} = 8\,\text{V}$$

$$V_{cd} = 4\,\text{V}$$

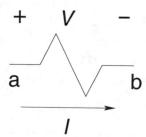

FIGURE 3.4: A resistor.

We have discovered some spaces using the rules that allow a circuit to function as a whole. These rules include the two Kirchhoff's laws and the rules that relate the current through a component to the voltage across it. We have found that the spaces we need to concentrate on are the nodes in the circuit. In the next section we introduce the simplest of all electrical components, the resistor, and apply what we have learned to solving circuits containing resistors.

3.3   THE RESISTOR

The simplest kind of connection analogous to a hollow pipe is something called a resistor. The symbol for the resistor is shown in Fig. 3.4. Resistors with a high-resistance value will let less current flow through it for a given voltage than a resistor with a low-resistance value. The most popular kinds of resistors are *linear* resistors. When we say a function is linear we mean that the function $f()$ or $g()$ corresponds to a line through the origin. In particular

$$V = IR \tag{3.18}$$

where R is the resistance measured in ohms[1] with symbol Ω. or

$$I = GV \tag{3.19}$$

---

[1] George Simon Ohm was born in Erlangen, France, on March 16, 1789. He published his observations on the relationship between voltage, current, and resistance in 1827. The work was not well received and Ohm was so upset that he resigned from his academic position.

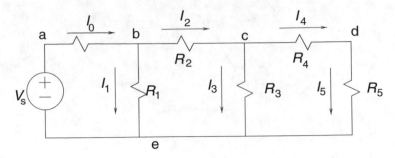

FIGURE 3.5: An example of a resistive circuit.

where $G = 1/R$ is called the conductance and is measured in siemens with a symbol S.[2] The law relating the current through a resistor to the voltage across it [Eq. (3.18)] is called Ohm's law. If we plot V versus I in these equations we can see that the plot is a straight line through the origin, i.e., the resistor is a linear component (our earlier declaration). *Note that this relationship is valid only when the current I flows from the assumed positive terminal of the resistor to the assumed negative terminal.*

3.4   RESISTIVE CIRCUITS

Let's examine how we would go about analyzing a circuit that consists of voltage sources and resistors. Consider the circuit shown in Fig. 3.5. Suppose we are asked to find all the voltages and currents in the circuit. Recall that if we know the function relating the voltage across a component to the current through the component and we know the voltages at each of the nodes with respect to a common reference node, we can easily find the currents through the components. Once we have identified the nodes our next task is to choose one of the nodes as the reference node. As we are going to find the voltage of all other nodes with reference to the reference node, it makes the algebra simpler if we select the node that has the largest number of components attached to it. In the case of the circuit shown in Fig. 3.5, it is node e.

---

[2] Because conductance is the inverse of resistance it is sometimes measures in mhos with the symbol being an upside down Ω.

Having declared node e to be our reference node, we go to each of the other nodes and based on the information available to us we do one of two things:

1. Determine the voltage at that node with respect to the reference node. OR

2. Write the current law at the node.

Node a: We begin with node a. From the circuit diagram in Fig. 3.5 it is clear that the potential difference between node a and node e is V_s volts, or

$$V_{ae} = V_s \qquad (3.20)$$

Node b: The potential difference between node b and node e is not immediately apparent so we write the current law at this node. Given the current directions shown in the circuit diagram we can see that we have one current (I_0) entering the node and two currents (I_1 and I_2) leaving the node. Therefore, according to the Kirchhoff's current law

$$I_0 = I_1 + I_2 \qquad (3.21)$$

Node c: Similarly at node c the potential difference between node c and node e is not immediately apparent, therefore, we write the current law at this node. We have the current I_2 entering the node and the currents I_3 and I_4 leaving the node. Therefore,

$$I_2 = I_3 + I_4 \qquad (3.22)$$

Node d: At node d we have current I_4 entering the node and the current I_5 leaving the node. Therefore, according to the current law

$$I_4 = I_5 \qquad (3.23)$$

The next step is to write each of these currents in terms of the voltages across the resistors. At this stage we need to be careful that the directions we have assigned to the currents agree with the assumed polarities. For example, $I_0 = V_{ab}/R_0$ and **not** V_{ba}/R_0. We can express each of the currents in terms of the voltages across the respective resistors using Ohm's law. Thus,

$$I_0 = \frac{V_{ab}}{R_0} \qquad I_1 = \frac{V_{be}}{R_1}$$

$$I_2 = \frac{V_{bc}}{R_2} \qquad I_3 = \frac{V_{ce}}{R_3}$$

$$I_4 = \frac{V_{cd}}{R_4} \qquad I_5 = \frac{V_{de}}{R_5}$$

Substituting these in Eqs. (3.20)–(3.22) we obtain

$$\frac{V_{ab}}{R_0} = \frac{V_{be}}{R_1} + \frac{V_{bc}}{R_2} \tag{3.24}$$

$$\frac{V_{bc}}{R_2} = \frac{V_{ce}}{R_3} + \frac{V_{cd}}{R_4} \tag{3.25}$$

$$\frac{V_{cd}}{R_4} = \frac{V_{de}}{R_5} \tag{3.26}$$

Now we express all voltages in terms of the voltage at each node with respect to the reference node. Thus we replace V_{ab} by $V_{ae} - V_{be}$, V_{bc} by $V_{be} - V_{ce}$, and V_{cd} by $V_{ce} - V_{de}$. Finally, noting that $V_{ae} = V_s$ we obtain the following equations:

$$\frac{V_s - V_{be}}{R_0} = \frac{V_{be}}{R_1} + \frac{V_{be} - V_{ce}}{R_2} \tag{3.27}$$

$$\frac{V_{be} - V_{ce}}{R_2} = \frac{V_{ce}}{R_3} + \frac{V_{ce} - V_{de}}{R_4} \tag{3.28}$$

$$\frac{V_{ce} - V_{de}}{R_4} = \frac{V_{de}}{R_5} \tag{3.29}$$

If we know the values of the resistors, the unknowns in these equations are V_{be}, V_{ce}, and V_{de}. We have three linear equations and three unknowns. Three equations with three unknowns is really not a big deal and we can easily solve this. Even if the problem gets larger there are a number of tools available for solving linear equations and we can easily obtain these voltages. Once we have the node voltages we can obtain any voltage or current we desire. For example, suppose we were asked for

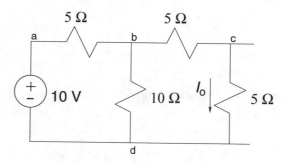

FIGURE 3.6: An example of a linear circuit.

the voltage V_{bd}. This is easily obtained as

$$V_{bd} = V_{be} - V_{de}$$

How would you find the current I_{ca} through the voltage source?

There are all kinds of variations we can have on this basic theme. Let's take a look at a few examples.

Example 3.4.1: Consider the circuit shown in Fig. 3.6. Suppose we are asked to find the current I_o. If we knew the voltage V_{cd} across the 5Ω resistor at the right, we could then obtain the current I_o using Ohm's law. We have four nodes a, b, c, and d. Let's pick node d to be the reference node. We go to each of the other nodes and either write the voltage at that node with respect to the reference node, or write the current law at that node.

At node a: $V_{ad} = 10\,\text{V}$
At node b: $I_{ab} = I_{bd} + I_{bc}$
At node c: $I_{bc} = I_{cd}$

Notice that each of the currents is the current through a resistor. Using Ohm's law we can write each of these currents as the voltage across the resistor divided by the resistance value.

At node b:

$$\frac{V_{ab}}{5} = \frac{V_{bd}}{10} + \frac{V_{bc}}{5}$$

At node c:

$$\frac{V_{bc}}{5} = \frac{V_{cd}}{5}$$

To simplify these equations we multiply the first one by 10 and the second one by 5.

$$2V_{ab} = V_{bd} + 2V_{bc} \tag{3.30}$$

$$V_{bc} = V_{cd} \tag{3.31}$$

We have two equations but four unknowns. To reduce the number of unknowns we write these voltages in terms of the node voltages. Substituting $V_{ab} = V_{ad} - V_{bd}$ and $V_{bc} = V_{bd} - V_{cd}$ and using the fact that $V_{ad} = 10\,V$ we obtain

$$2(10 - V_{bd}) = V_{bd} + 2(V_{bd} - V_{cd}) \tag{3.32}$$

$$V_{bd} - V_{cd} = V_{cd} \tag{3.33}$$

We have reduced the number of unknowns to two: V_{bd} and V_{cd}. With two equations and two unknowns we can solve for the unknowns and we obtain

$$V_{cd} = 2.5\,V$$

From which we get

$$I_o = \frac{2.5}{5} = 0.5\,A$$

Suppose we also wanted to find the voltage V_{ac} and the current through the voltage source I_{da}.

$$V_{ac} = V_{ad} - V_{cd}$$

Therefore,

$$V_{ac} = 10 - 2.5 = 7.5\,V$$

To find I_{da} we need to write the current law at node d.

$$I_{da} = I_{bd} + I_{cd}$$

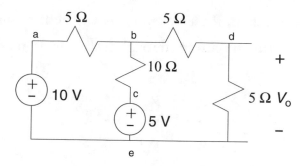

FIGURE 3.7: Another example of a linear circuit.

The current I_{cd} is the same as I_o, which we have already found to be 0.5 A. The currents I_{bd} can be obtained using Ohm's law if we know V_{bd}. We can obtain V_{bd} from Eq. (3.33).

$$V_{bd} = 2V_{cd} = 5\,\text{V}$$

Therefore,

$$I_{cd} = \frac{5}{10} = 0.5\,\text{A}$$

and

$$I_{da} = 0.5 + 0.5 = 1\,\text{A}$$

Example 3.4.2: Let's look at a slightly different problem as shown in Fig. 3.7. At first glance this looks more complicated. There are five nodes instead of four. However, if we look at this a bit closer we notice that (assuming node e is the reference node), as in the case of the circuit in Fig. 3.6, we again have only two nodes, node b and node d, for which we do not know the node voltages. The voltage at node a with respect to node e is 10 V, while the voltage of node c with respect to node e is 5 V. In order to find the voltages at nodes b and d with respect to node e we write the current law at these nodes.

At node a: $V_{ae} = 10\,\text{V}$

At node b: $I_{ab} = I_{bc} + I_{bd}$

At node c: $V_{ce} = 5\,\text{V}$

At node d: $I_{bd} = I_{de}$

Each of the currents in these equations is a current through a resistor. We can use Ohm's law to write the current through each resistor in terms of the voltage across that resistor.

$$\frac{V_{ab}}{5} = \frac{V_{bc}}{10} + \frac{V_{bd}}{5} \tag{3.34}$$

$$\frac{V_{bd}}{5} = \frac{V_{de}}{5} \tag{3.35}$$

Multiplying the top equation by 10 and the bottom equation by 5 we get

$$2V_{ab} = V_{bc} + 2V_{bd} \tag{3.36}$$

$$V_{bd} = V_{de} \tag{3.37}$$

Writing these voltages in terms of the node voltages and substituting $V_{ae} = 10\,\text{V}$ and $V_{ce} = 5\,\text{V}$ we get

$$2(10 - V_{be}) = V_{be} - 5 + 2(V_{be} - V_{de}) \tag{3.38}$$

$$V_{be} - V_{de} = V_{de} \tag{3.39}$$

Solving for V_{de} (which is the same as V_o) we obtain

$$V_o = V_{de} = 3.125\,\text{V}$$

In the previous examples, we have been using resistors and voltage sources. However, the Kirchhoff's laws are not restricted to these components.

Example 3.4.3: Consider the circuit shown in Fig. 3.8. This looks very much like the circuit in Fig. 3.6 so we can write the current law at nodes b and c.

At node a: $V_{ad} = 10\,\text{V}$

At node b: $I_{ab} = I_{bd} + I_{bc}$

At node c: $I_{bc} = I_{cd}$

At this point we run into a problem. The component between nodes b and c is not a resistor so we cannot write the current I_{bc} in terms V_{bc} using Ohm's law. In

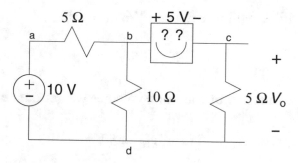

FIGURE 3.8: Yet another example of a linear circuit.

fact we do not know anything about this component except that the voltage across it is 5 V. Let's use what information we have leaving I_{bc} alone for now. Writing the currents through resistors in terms of voltages across resistors we get

$$\frac{V_{ab}}{5} = \frac{V_{bd}}{10} + I_{bc} \qquad (3.40)$$

$$I_{bc} = \frac{V_{cd}}{5} \qquad (3.41)$$

We can substitute the value of I_{bc} from Eq. (3.41) into Eq. (3.40) to obtain

$$\frac{V_{ab}}{5} = \frac{V_{bd}}{10} + \frac{V_{cd}}{5}$$

Simplifying this equation by multiplying it by 10

$$2V_{ab} = V_{bd} + 2V_{cd}$$

and writing the voltages in terms of the node voltages we obtain

$$2(V_{ad} - V_{bd}) = V_{bd} + 2V_{cd}$$

Substituting $V_{ad} = 10$ we have

$$20 - 2V_{bd} = V_{bd} + 2V_{cd}$$

We have one equation and two unknowns, V_{bd} and V_{cd}. If we had no information about the mystery component, we would be stumped at this point. Fortunately, we do have some additional information. We know that

$$V_{bc} = 5\,V$$

Writing this voltage in terms of the node voltages we get our second equation

$$V_{bd} - V_{cd} = 5\,V$$

With two equations we can solve for the two unknowns and we get

$$V_o = V_{cd} = 1\,V$$

Example 3.4.4: Let's see if we can use what we have learned to this point to design a circuit. Suppose we have need for a 5 V source, perhaps to power a logic circuit, but we only have a 9 V battery available. We know that if we connect a sequence of resistors to the 9 V battery we will get differing voltages across the resistors. Therefore, a possible solution to our problem would be a circuit of the form shown in Fig. 3.9. So what should R_1 and R_2 be? Let's start with what we know and write the current law at node b.

$$I_{ab} = I_{bc}$$

Picking node c to be the reference node we can see that $V_a = 9\,V$ and $V_b = 5\,V$. Therefore, we can write the current law as

$$\frac{9 - 5}{R_1} = \frac{5}{R_2}$$

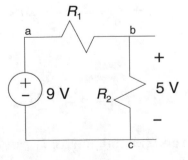

FIGURE 3.9: voltage divider.

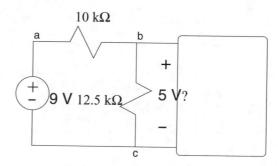

FIGURE 3.10: Still 5 volts?

We have one equation and two unknowns. When we have fewer equations than unknowns we can pick a value for one of the unknowns and solve for the other. Let's pick R_1 to have a value of 10 kΩ (we could have picked any other value as well). Solving for R_2 we obtain $R_2 = 12.5$ kΩ.

Substituting these values for the resistors in Fig. 3.9 we will obtain 5 V between nodes b and c. However, will the voltage still be 5 V after we have connected the system that needs the 5 V as shown in Fig. 3.10?

What do you think could go wrong? What would we need to do to ensure that the voltage stays at 5 V?

3.5   SUMMARY

In this chapter we have introduced our first component, the resistor, and its component rule, the Ohm's law ($V = IR$). Using the two circuit laws KVL and KCL and the component rule we have shown that we can analyze any resistive network.

3.6   PROJECTS AND PROBLEMS

1. In the circuit shown in Fig. 3.11 what do you think will happen when we open the switch? Why? What is the function of the resistor? What practical use can you make of this circuit?

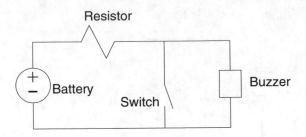

2. In the circuit shown in Fig. 3.12 what will happen as the switches are opened? Why?

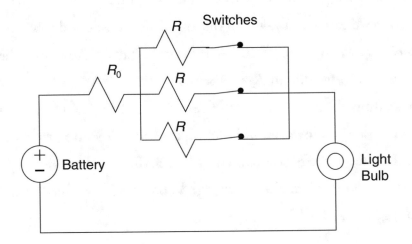

3. In the circuit shown in Fig. 3.13 what will happen when the switches are opened? Why?

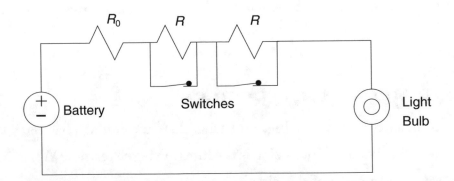

4. In the circuit shown in Fig. 3.14 pick R_1, R_2, R_3 and R_o so that the light does not turn on.

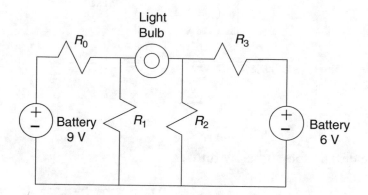

5. In the circuit below the components A, B, and C, have the following current–voltage relationships.

$$V_A = 3I_A + 12$$

$$V_B = 4I_B$$

$$V_C = 2I_C$$

Find V.

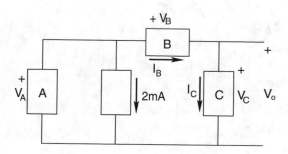

6. In the circuit shown below find the voltage across the resistor labeled R.

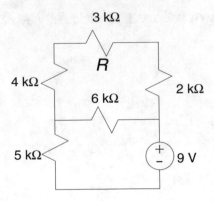

7. In the circuit shown below find V.

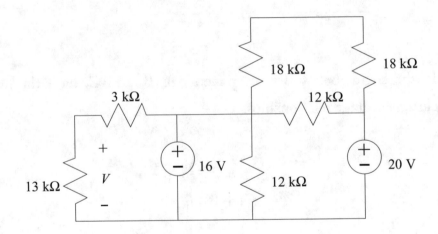

8. In the circuit shown below find the voltage V_o.

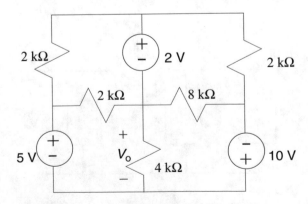

9. In the circuit shown below find I_o.

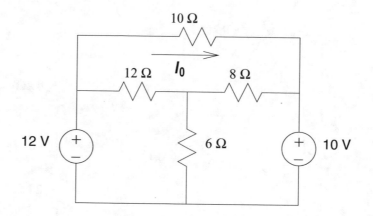

10. In the circuit shown below find the voltage across the $2\,\Omega$ resistor. The component labeled A has the voltage current relationship

$$V_A = 10.5 + 7\,I_A$$

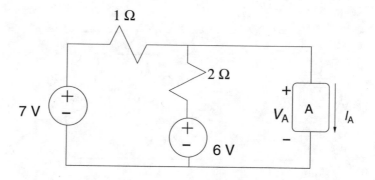

11. In the following circuit find V_o.

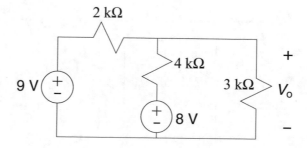

12. In the following circuit find V_o, if $V_s = 12\,\text{V}$.

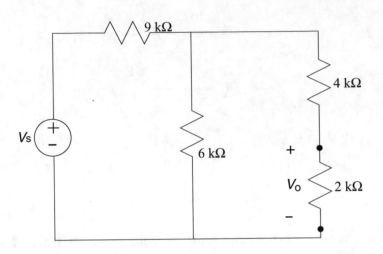

13. In the following circuit find I_o.

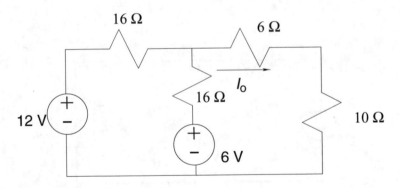

14. In the following circuit find I_o.

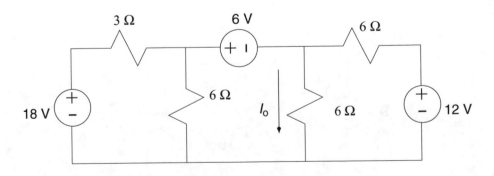

15. In the circuit shown below find I_o.

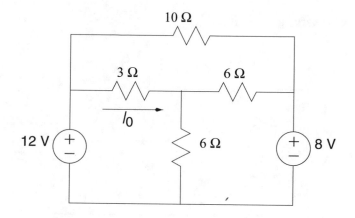

16. In the following circuit find V_o.

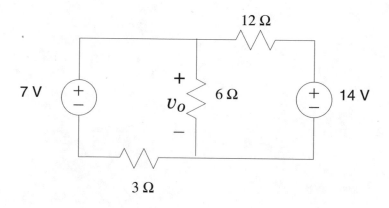

17. In the following circuit find V_o.

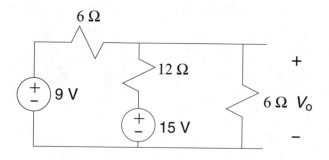

18. In the following circuit find I_o.

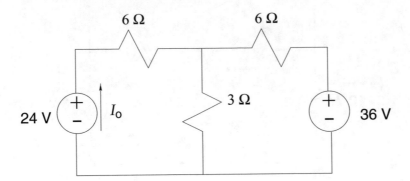

C H A P T E R   4

Capacitors

4.1 OVERVIEW

In this chapter we look at another component—the capacitor—and its use in a very common application of electrical circuits. We will begin by introducing the capacitor, then describe the concepts of frequency and filtering.

4.2 INTRODUCTION

The capacitor is made up of a nonconducting or *dielectric* material sandwiched between two *conductors*. A conductor is a material in which the valence electrons are loosely bound and therefore can be used to transport electric charge. Standard conductors include copper and iron. An insulator is a material in which the valence electrons are tightly bound and therefore do not transport charge. Standard dielectric material includes air, paper, Mylar, mica, glass, or ceramics.

Consider the simple circuit shown in Fig. 4.1. The battery creates a potential difference between its positive and negative terminals. If there was a connection between these two terminals there would be a flow of current (remember we assume that the current is a flow of positive charges). However, the presence of the capacitor prevents a direct flow between the two terminals. Initially, there is a flow of positive charge from the positive terminal of the battery. The charge collects on plate A of the capacitor. The charge cannot flow through the dielectric. However, it can exert a repelling force on the positive charge on the other conductor plate. The charge repelled from plate B flows to the negative terminal of the battery and there is a momentary flow of current in the circuit. After a while, there is an accumulation

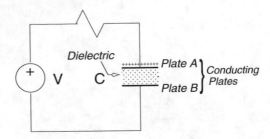

FIGURE 4.1: A capacitor.

of positive charge on plate A and an accumulation of negative charge on plate B. This creates a potential difference, which tries to push the current flow in the opposite direction. When the potential difference generated by the battery and the potential difference generated by the capacitor plates are equal, no more current flows through the circuit.

We can see that in the beginning of this process the voltage across the plates of the capacitor is zero. Therefore we get a substantial amount of current flow. As the voltage across the capacitor increases, this opposes the current flow until when the voltage across the capacitor reaches its maximum value the current flow is reduced to zero. Mathematically the voltage across a capacitor V_c and the current through it I_c are related by a derivative operation

$$I_c = C \frac{dV_c}{dt} \tag{4.1}$$

where C is the capacitance and is a measure of how much charge can be stored on the plates for a given potential difference. This is the component rule for the capacitor just like Ohm's law is the component rule for resistors.

In part this equation says what we have already observed. That is, when the voltage across the capacitor becomes constant the current through it is zero. (The derivative of a constant is zero.) However, it also says something very important that we may have missed in our previous observations: in order for the voltage to change instantaneously, we need an infinite amount of current to flow through the capacitor. This is because an instantaneous change in $V(t)$ would mean that the slope of at that point, or the derivative, $V(t)$ is infinite. Because this is not feasible in practice, the voltage across a capacitor cannot change instantaneously. This gives the circuit a connection with its immediate past. In other words the circuit "remembers" what the voltage was in the immediate past. If we see a circuit containing only resistors and the voltage across a particular resistor is 5 V, this tells us nothing about what the voltage was an instant ago. However, if the voltage across a capacitor is 5 V we know that right before our observation the voltage across it was about 5 V and right after our observation the voltage will be about 5 V. If we did try to force a sudden change in the voltage, for example, by shorting out the terminals, we would get a large (and dangerous) surge in the current. Therefore, a capacitor can be used as a memory element in a circuit.

The other observation we can make, based on this equation, is that the current through the capacitor depends on how the voltage varies with time. If the voltage varies rapidly we get a large current flow, and if it varies slowly (or not at all) we get a small (or no) current flow. How fast or how slow a signal varies with time can be expressed in terms of frequency. The behavior of the capacitor is frequency dependent. We can view the capacitor as a resistor whose resistance changes depending on how fast the voltage across it changes. If the voltage changes rapidly, the capacitor acts like a resistor with very low resistance resulting in a high value for the current through it. If the voltage changes slowly, the capacitor acts as a resistor with very high resistance. If the voltage is constant (after an initial transient), the capacitor acts as an open circuit with no current through it.

The concept of frequency is very important in the study of electrical engineering, as well as many other fields, and so we will take a brief detour to examine it a bit more closely.

4.3 FUNCTIONS OF TIME AND THE CONCEPT OF FREQUENCY

We often design circuits to handle signals that change with time. These signals come in all shapes and sizes. We can have signals that vary in a very regular fashion, such as the voltage on a power line, or in a very irregular fashion, as in the case of voice

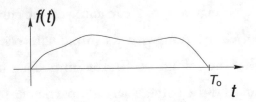

FIGURE 4.2: A function of time.

signals over telephone lines or the atmosphere. We would like to have a framework in which to represent all these signals. In the early part of the nineteenth century, in order to solve equations describing the dissipation of heat, Jean B.J. Fourier came up with such a framework. He showed that all periodic signals could be written in terms of sines and cosines. Many signals that we are interested in, such as audio and video signals, are not periodic. However, as these signals are nonzero only over a finite period of time, we can convert them into a periodic signal by repeating the signal. For example, consider the signal $f(t)$ shown in Fig. 4.2. By repeating it as shown in Fig. 4.3 we can generate a periodic signal that is identical to the original signal in the interval $[0, T_0]$. Thus the method of Fourier allows us to represent a variety of signals in a unified manner.

A sinusoidal signal is a signal of the form

$$V(t) = A\cos(2\pi ft + \theta) \tag{4.2}$$

where f is the *frequency* of the sinusoid and θ is the *phase*. The frequency of a sinusoid is measured in cycles per second or Hertz[1] (Hz). A frequency of n Hz means that it takes 1 s for the sinusoid to complete n cycles. One cycle of the sinusoid $V(t) = 10\cos(2\pi \times 10t)$ is shown in Fig. 4.4. The amount of time taken by the sinusoid to complete one cycle is called the *period* of the sinusoid, and is usually denoted by T. We can see that the period is simply the reciprocal of the frequency:

$$T = 1/f$$

---

[1] Heinrich Rudolf Hertz was born on February 22, 1857, in Hamburg, Germany, and studied engineering and physics at Munich and Berlin. While he was a professor at the technical school in Karlsruhe, he conducted a set of experiments showing the propagation of electromagnetic waves (he called them electric rays). He was not yet 37 years old when he died on January 1, 1894.

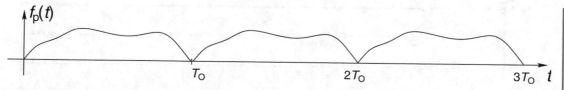

FIGURE 4.3: A periodic extension of $f(t)$.

One way of physically interpreting frequency is to view it as a measure of how fast the signal changes. A frequency of 2 Hz means that it takes the sinusoid half a second to complete one cycle; a frequency of 1000 Hz means that completing a cycle takes only one thousandth of a second.

Fourier showed that any periodic signal $f(t)$ with period T_o can be written as

$$f(t) = a_0 + \sum_{n=1}^{\infty} a_n \cos(2\pi n f_o t) + \sum_{n=1}^{\infty} b_n \sin(2\pi n f_o t) \qquad (4.3)$$

where a_n and b_n are called the Fourier coefficients and correspond to the components of $f(t)$ that change with frequency $n f_o$. For signals that change slowly the coefficients a_n and b_n will be close to zero for larger values of n. Such signals

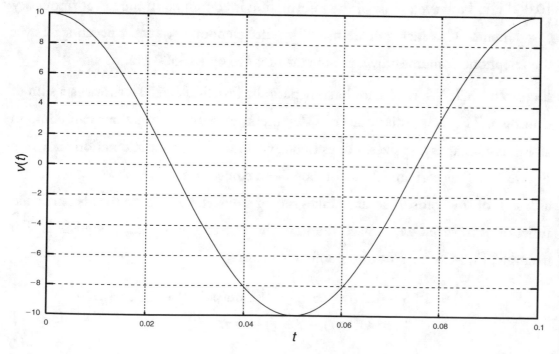

FIGURE 4.4: One cycle of sinusoid with frequency 10 Hz.

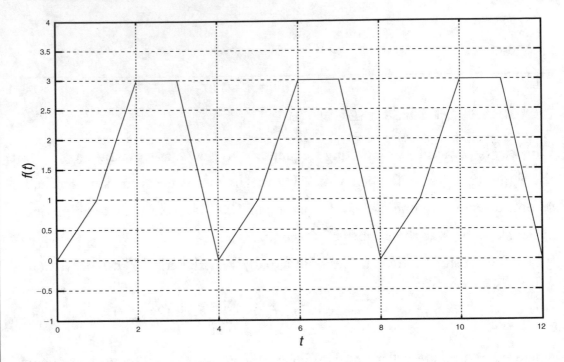

FIGURE 4.5: A periodic function.

are called low-frequency signals. The electrical signal that powers your house is a sinusoid at 60 Hz. The human voice contains frequency components of up to 10,000 Hz. However, most of the information is contained in the lower frequency components. This fact is made use of by the telephone system. Speech sent over the telephone contains only components of voice up to 3600 Hz.

Example 4.3.1: We have said that any periodic function can be written as a sum of sinusoids. This is a rather powerful statement, as periodic functions can come in all kinds of shapes and sizes. For example, consider the periodic function shown in Figure 4.5. This certainly does not look like a sinusoid. However, we can write this as a sum of sinusoids. It actually takes lots of sinusoids to add up to this particular function. However, we can see the trend by just adding up a few sinusoids. For example, Fig. 4.6 is a plot of the following sum

$$f(t) = 1.75 - 1.2\cos(0.5\pi t) - .81\sin(0.5\pi t) - 0.2\cos(\pi t)$$
$$- .12\cos(1.5\pi t) - .09\sin(1.5\pi t)$$

superimposed on the plot of Fig. 4.5.

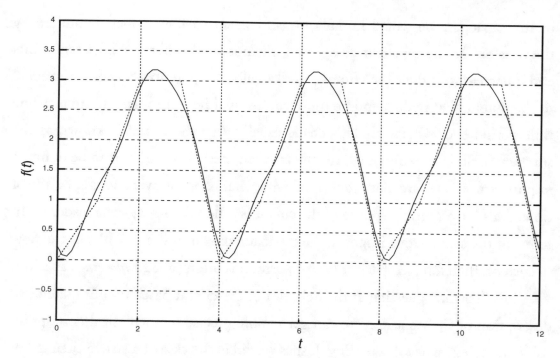

FIGURE 4.6: A sum of sinusoids.

In later courses you will learn how to decompose a signal into its frequency components. This brief introduction was just to get you thinking about the fact that there is more than one way of looking at a function.

4.4   CIRCUITS WITH RESISTORS AND CAPACITORS

Let us build a simple circuit with a voltage source, a resistor, and a capacitor. A possible setup is shown in Fig. 4.7. Given what we have described about the functioning

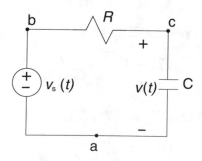

FIGURE 4.7: A simple RC circuit.

of the capacitor, we would initially expect current to flow, which is limited by the resistor. As the capacitor charges up, the potential difference between points and c decreases as does the current, until finally the potential at node c is equal to the potential at node d and no more current will flow through the circuit. The time period over which this happens is usually very short, and is known as the *transient* period. In many situations the transient period is too short to be of much importance, and we are more concerned with what happens over a longer period of time. This longer term response of the circuit is called the *steady-state* response. In terms of the steady-state response the capacitor shown in Fig. 4.7 blocks the flow of current, therefore, a capacitor in this position is often called a *blocking* capacitor.

Rather than guessing at the behavior of the circuit based on our qualitative description of the capacitor, we can get a more precise mathematical description using the rules we have been using. Just as we did in the case of resistive circuits, we can obtain all the voltages and currents in the system by writing the current law at each node whose voltage we do not know. Let's pick node a to be our reference node.

At node b: $V_{ba} = V_s(t)$
At node c: $I_{bc} = I_{ca}$

I_{ba} is the current through a resistor and I_{ca} is the current through the capacitor. Therefore, in order to write the current through these components in terms of the voltage across the components we need to use the corresponding component rule: Ohm's law for the resistor and Eq. (4.1) for the capacitor.

$$\frac{V_{bc}}{R} = C\frac{d}{dt}V_{ca}$$

Writing V_{bc} in terms of the node voltages

$$\frac{V_{ba} - V_{ca}}{R} = C\frac{d}{dt}V_{ca}$$

From Fig. 4.7, $V_{ba} = V_s(t)$ and $V_{ca} = V(t)$. Therefore,

$$\frac{V_s(t) - V(t)}{R} = C\frac{d}{dt}V_s(t)$$

Simplifying this we obtain

$$\frac{dV(t)}{dt} + \frac{1}{RC}V(t) = V_s(t)$$

What we have obtained is a *differential equation*. Many of you may not have studied differential equations yet. But you will, and at least some of you (I hope many) will find it a fascinating topic. Differential equations provide us with a compact way of describing the behavior, or modeling, of physical systems. Unfortunately, the study of differential equations is beyond the scope of this book. So, instead of trying to solve differential equations we will use two approximations to the behavior of capacitors. When $V_s(t)$ is a constant, we will assume that in steady state the capacitor is an open circuit. When $V_s(t)$ is a high-frequency signal, we will assume that in steady state the capacitor is a short circuit. In other words the resistance of the capacitor goes from infinity at zero frequency to zero at high frequencies. Thus the capacitor behaves as a frequency-dependent resistor.

Given these approximations let's find the voltage $V(t)$ in the circuit of Fig. 4.7 for the two cases:

1. When $V_s(t)$ is a constant

$$V_s(t) = V$$

    according to our approximation our circuit looks as shown in Fig. 4.8. Using the voltage law

$$V(t) = V_{ca} = V_{cb} + V_{ba}$$

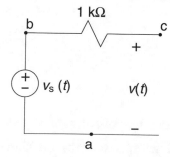

FIGURE 4.8: The open circuit approximation.

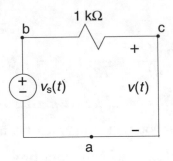

FIGURE 4.9: The short circuit approximation.

As there is no current flowing through the resistor $V_{cb} = 0$ and $V(t) = V_{ba} = V$.

2. When $V_s(t)$ is a high-frequency signal, the capacitor acts as a short circuit as shown in Fig. 4.9. In this case $V(t)$ is zero.

We can use this property of the capacitor to block low-frequency signals from a circuit as shown in Fig. 4.10(a), or as a shunt to divert high-frequency signals from a circuit as shown in Fig. 4.10(b).

The capacitor blocks current flow when the voltage across it is constant and acts as a short circuit when the voltage across it is changing rapidly. Another way of looking at this is to view the capacitor as a resistor whose resistance value changes with frequency. Thus the capacitor is like a resistor with a high-resistance value when the voltage across it changes slowly and a resistor with a low-resistance value when the voltage across it changes rapidly. This change in resistance is a continuous progression: as we change the rate of change of the voltage, or the frequency of

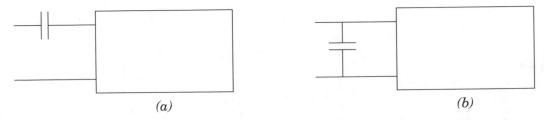

(a)   (b)

FIGURE 4.10: (a) The capacitor blocks constant signals (signals with zero frequency) from the rest of the circuit. (b) The capacitor prevents high-frequency signals from getting to the circuit.

the voltage signal, the "resistance" of the capacitor decreases. We make use of this frequency dependent resistance property of the capacitor to build *filters*. Filters selectively let through or attenuate signals depending on their frequency. Filters are part of all communication devices including radios, televisions, and cell phones. If you have an audio system it probably has an equalizer. This is just a bank of filters that amplifies or attenuates specific frequency components of the audio signal.

When a filter blocks out or attenuates all signals with frequencies above a *cut-off frequency*, and lets through signals with frequencies below the cutoff frequency, it is called a *low-pass filter*. You might use a filter like this to block out hissing noise from a voice signal. Human voice does not have many components above around 7000 Hz, while hissing noise has many components above that value. So a low-pass filter with a cutoff at 7000 Hz would do much to rid the voice signal of hiss.

A filter that blocks out or attenuates signals with frequencies below a cutoff frequency while letting through signals with frequencies above a cutoff frequency is called a *high-pass filter*. You might use such a filter to remove the 60 Hz hum due to the electrical wiring that might intrude on a voice signal. Again, human voice has almost no components at frequencies that low so signals at that frequency can be blocked out without losing any voice information.

Finally, we can design filters that let through or attenuate a band of frequencies. These are called *band pass filters* or *band stop filters*.

In order to see what is meant by "letting a signal through" or "blocking a signal" let's analyze a couple of simple filters. Consider the circuit shown in Fig. 4.11

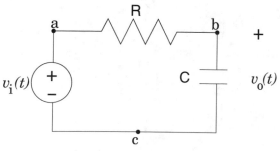

FIGURE 4.11: Low pass filter.

where $V_i(t)$ is the input voltage and $V_o(t)$ is the output voltage. When $V_i(t)$ is constant (zero frequency) the capacitor acts as an open circuit and there is no current through the resistor. Therefore,

$$V_o(t) = V_{bc} = V_{ba} + V_{ac}$$

The voltage V_{ba} is the voltage across the resistor and V_{ac} is the input voltage. If the current through the resistor is zero the voltage across it is also zero. Therefore,

$$V_o(t) = V_i(t)$$

the input is passed through to the output. At high frequencies we can approximate the capacitor with a short circuit. Because the voltage across a short circuit is zero, the output voltage $V_o(t)$ is zero regardless of the input voltage. In other words the input voltage cannot be perceived, or is blocked, from the output. This is a rather crude analysis as we are only looking at the extreme conditions of zero frequency and very high frequency. However, even with this we can see that this circuit will let through the low-frequency signal and block the high-frequency signals.

If we now swap the positions of the resistor and capacitor, as shown in Fig. 4.12, we have a different filter. In this configuration, the output voltage is the voltage across the resistor. At low frequencies the capacitor reduces or blocks the current through the resistor. According to Ohm's law this means the voltage across the resistor will be low, or zero for the case where the current is completely blocked.

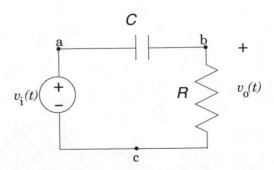

FIGURE 4.12: High pass filter.

Thus the input voltage is not reflected at the output. When the frequency of the input signal is high the capacitor acts as a short circuit and the voltage across the resistor is the same as the input voltage. Thus the circuit lets through high-frequency signals while blocking low-frequency signals. Therefore, in this configuration the circuit is a high-pass filter.

4.5 SUMMARY

In this chapter we have introduced our second circuit component, the capacitor, and its component rule, $I = C(dV/dt)$. Unlike the resistor, the operation of the capacitor is connected to how the voltage across it and the current through it changes with time. The two circuit laws KVL and KCL along with the component rules for the resistor and capacitor are sufficient to analyze any circuit made up of these components. However, the equations we would have to solve are differential equations. As most readers may not have learned how to use differential equations at this stage we have not tried to analyze circuits of any complexity. Instead, we have looked at two approximations that can be used for the cases where we have slowly varying voltages and currents, and the cases where we have rapidly varying voltages and currents.

4.6 PROJECTS AND PROBLEMS

1. In the circuit shown in Fig. 4.11 the input voltage $V_i(t)$ is the square wave shown in Fig. 4.13. Sketch the output voltage $V_o(t)$.

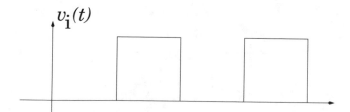

FIGURE 4.13: Plot of input voltage $V_i(t)$.

2. In the following circuit the input voltage $V_i(t)$ is the square wave shown in Fig. 4.13. Sketch the output voltage $V_o(t)$.

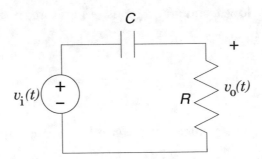

3. Shown below is a plot of the voltage across a capacitor and a plot of the corresponding current through the capacitor. Which plot (the solid or dashed) is a plot of the voltage and which is a plot of the current? Explain your choices.

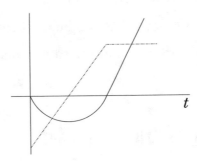

4. The following graph is either that of voltage across a capacitor or current through a capacitor. Which one is it? Justify your answer.

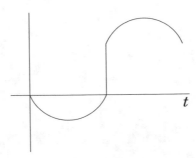

CHAPTER 5

Diode Circuits

5.1 OVERVIEW

In this chapter we look at our first nonlinear component, the diode. We will look at some popular applications of diodes and look at one way of analyzing circuits that contain diodes.

5.2 INTRODUCTION

Both the resistor and the capacitor are *linear* elements. Suppose the current I through a component and the voltage V across it are related as

$$I = g(V)$$

A linear component will have the following two properties.

1. **Homogeneity:** If we have two voltages V_1 and V_2 that correspond to currents $I_1 = g(V_1)$, and $I_2 = g(V_2)$, then the current corresponding to the sum of two voltages is simply the sum of the two currents. In other words

$$g(V_1 + V_2) = I_1 + I_2$$

2. **Scaling:** If we scale the voltage by a factor α the current is also scaled by the same factor.

$$g(\alpha V) = \alpha g(V) = \alpha I$$

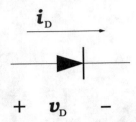

FIGURE 5.1: Symbol for diode.

Both the resistor and the capacitor have component rules that satisfy these two properties. However, not all components have component rules that satisfy this property. One of the most common nonlinear components is the *diode*. The diode is represented by the symbol shown in Fig. 5.1.

The component rule for the diode, or the relationship between the current through the diode, I_D, and the voltage across it, V_D, is rather complex. It is given by

$$I_D = I_o(e^{\alpha V_D} - 1) \tag{5.1}$$

where α is a function of a number of things including the material used to construct the diode and the temperature. An example of a graph of current versus voltage (*I–V* curve) for a diode is shown in Fig. 5.2.

Because of this somewhat complex relationship between current and voltage, circuits containing diodes may be more difficult to solve than circuits containing resistors and capacitors. The Kirchhoff's laws can still be used to come up with the requisite number of equations, but these equations will involve transcendental expressions, which make them difficult to solve in the standard manner. Later in this chapter we will introduce a graphical approach that can be used to solve complex equations. However, what we usually do when we have diodes in a circuit is approximate their behavior.

Two approximations to the *I–V* curve are shown in Fig. 5.3. Consider the approximation shown in Fig. 5.3(b). According to this approximation as long as the voltage across the diode is less than a threshold there is no current flowing through

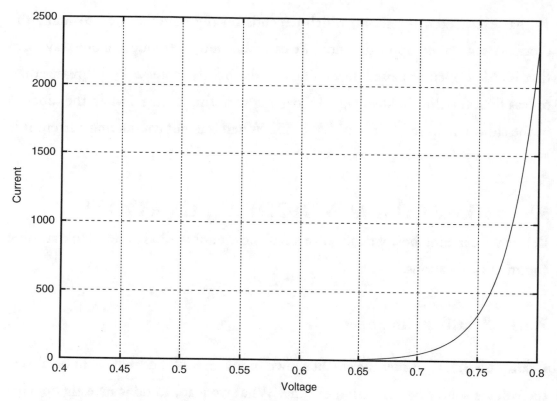

FIGURE 5.2: Current–voltage characteristic for a diode.

the diode. The diode acts as an open circuit. However, as soon as the voltage across the diode reaches the threshold, the diode acts as a short circuit. The threshold shown in the figure is about 0.7 V, which is common for a popular type of diode. As indicated by the symbol for the diode shown in Fig. 5.1, the diode allows current flow only in the direction of the arrow. The current can flow only after the voltage

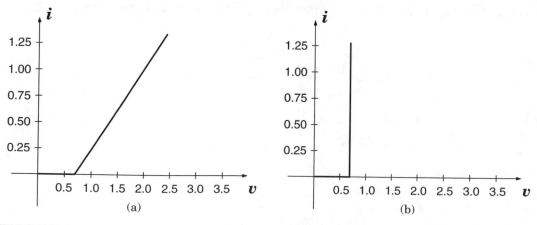

FIGURE 5.3: Approximations to the current–voltage characteristic for a diode.

V_D has exceeded a threshold, which depends on the material used to make the diode. Based on this approximation, we can then view the diode as a one way valve that is either open or closed depending on whether the voltage V_D is greater than or less than the threshold voltage. In keeping with this analogy, when the diode is in the short circuit mode it is said to be *on*. When it is not conducting current it is said to be *off*.

5.3   SOME COMMON DIODE APPLICATIONS

Before we examine how we can solve circuits containing diodes, let's look at some common applications.

5.3.1   Rectifier Circuits

Example 5.3.1: Consider the circuit shown in Fig. 5.4. Notice that we have labeled the voltage source by a function of time. What we want to do is investigate what happens to the voltage across the resistor V_R when the voltage changes above and below the threshold value of 0.7 V. In order to do so we use the sinusoidal signal shown in Fig. 5.5. Fig. 5.5 tells us what the voltage would be when measured at different times. For example, if we measured the voltage V_{ac} at any time before $t = t_0$ the voltage would be less than 0.7 V. What does that mean for V_R? Looking at the circuit we can see that as long as the voltage across the diode is less than 0.7 V there will be no current flowing through the diode and, therefore, no current

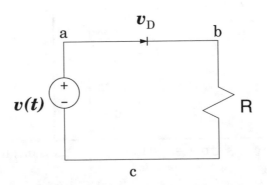

FIGURE 5.4: A simple diode circuit.

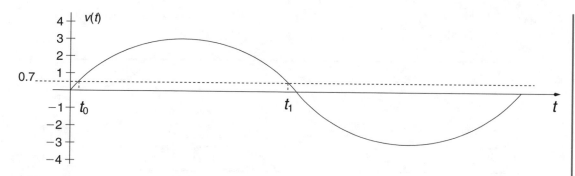

FIGURE 5.5: Input to the diode circuit.

flowing through the resistor. If the current through the resistor has a value of 0, then, by Ohm's law ($V = IR$), the voltage across the resistor or V_R will also be 0. We can see from the circuit that if the voltage difference generated by the voltage source between nodes a and c is less than 0.7 V, then the voltage difference between nodes a and b will also be less than 0.7 V. Therefore, for $t < t_0$ the voltage across the resistor V_R is 0.

$$V_R = \begin{cases} 0 & t < t_0 \\ ? & t > t_0 \end{cases}$$

Looking at Fig. 5.5 we see that the voltage is also less than 0.7 V for $t > t_1$. Using the same arguments as before the voltage across the resistor for $t > t_1$ will also be 0. Thus, our expression for V_R becomes

$$V_R = \begin{cases} 0 & t < t_0 \\ ? & t_0 < t < t_1 \\ 0 & t > t_1 \end{cases}$$

To find what happens between t_0 and t_1 let's use the Kirchhoff's voltage law. From the voltage law we know that

$$V_{ac} = V_{ab} + V_{bc}$$

Noting that $V_{ac} = V(t)$ and $V_{bc} = V_R$

$$V_R = V(t) - V_{ab}$$

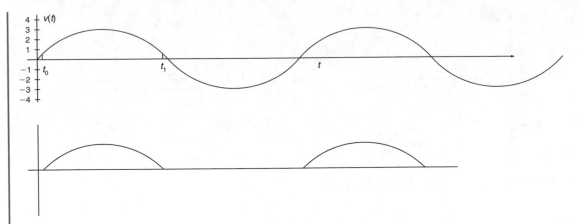

FIGURE 5.6: Input and output of half-wave rectifier.

Looking at Fig. 5.3(b) we can see that for the diode to continue to conduct, the voltage across it has to remain at the threshold. Therefore, for t between t_0 and t_1, $V_{ab} = 0.7$ V, and

$$V_R = \begin{cases} 0 & t < t_0 \\ V(t) - 0.7 & t_0 < t < t_1 \\ 0 & t > t_1 \end{cases}$$

The circuit we have described above is called a *rectifier*. It allows us to take a signal that has both positive and negative fluctuations and convert it to a signal that never turns negative. Because it does so by only letting the positive signal appear at the resistor and suppressing the negative signal it is called a half-wave rectifier. An example of an input and output of a half-wave rectifier is shown in Fig. 5.6.

The full-wave rectifier not only lets the positive signal through but also converts the negative signal into a positive signal. The circuit of a possible implementation of a full-wave rectifier (known as a bridge rectifier) is shown in Fig. 5.7. When the magnitude of the voltage is less than the diode threshold it is easy to verify that no current will flow through the resistor.

Let's look at the cases where the magnitude of the voltage is greater than the threshold. It is easiest to see how the full-wave rectifier works by treating the diode as a one-way valve. When the voltage $V(t)$ is positive, the current flows through the diode between nodes a and b. Then it flows through the resistor from node b to node c, and then through the diode between nodes c and d to complete the circuit. The

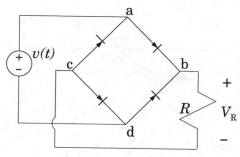

FIGURE 5.7: A full-wave rectifier.

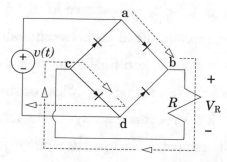

FIGURE 5.8: The path of the current through the full-wave rectifier during the positive cycle of the voltage $V(t)$.

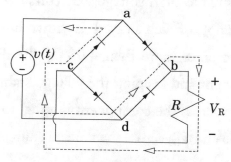

FIGURE 5.9: The path of the current through the full-wave rectifier during the negative cycle of the voltage $V(t)$.

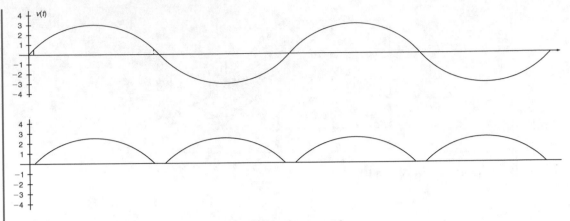

FIGURE 5.10: Input and output of a full-wave rectifier.

path of the current in the positive cycle is shown in Fig. 5.8. When the voltage is negative, the current flows through the diode between nodes d and b, then through the resistor from node b to node c, and finally through the diode between nodes c and a as shown in Fig. 5.9. Notice that regardless of the polarity of the voltage source, the current through the resistor is always in the same direction. Therefore, given the assigned polarity the voltage across the resistor V_R is always positive. An example of an input and output for a full-wave rectifier is shown in Fig. 5.10.

5.3.2 Diode Clamp

Another application of the diode is as a safety valve when we have components that should not be subjected to high voltages. Circuits that perform this function are called *clamps*. An example of such a circuit is shown in Fig. 5.11. In this circuit whenever the voltage becomes more than $+0.7$ V the diode diverts additional current from the component and keeping the voltage stabilized. To see this, note that V_{ac} the voltage drop across the component is equal to the sum of the voltage drop V_{ab} across the diode and V_{bc} across the voltage source:

$$V_{ac} = V_{ab} + V_{bc}$$

The voltage across the diode can be at most 0.7 V and the voltage at the source is V_o. Therefore,

$$V_{ac} \leq 0.7 + V_o$$

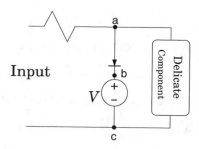

FIGURE 5.11: A diode clamp.

5.3.3 Voltage Limiter

This clamping property of the diode can also be used to limit the peak-to-peak swing of a voltage signal. A circuit that does that is called a *limiter*. The circuit diagram of a limiter is shown in Fig. 5.12. If the input voltage goes above 0.7 V or below −0.7 V one or the other of the diodes will begin to conduct and hold the voltage at ±0.7 V. Think about how this circuit could be used to generate a square wave.

5.4   SOLVING CIRCUITS CONTAINING DIODES

We have been looking at circuits where it is reasonably obvious when the diode is on or off. What do we do when we are not that sure about the condition of the diode? One approach that is relatively simple is to initially assume that the diode is on. This means that we assume that the voltage across the diode is 0.7 V and the current flows in the appropriate direction. We then solve the circuit using the approach we used previously. Then we check to see whether our assumption was

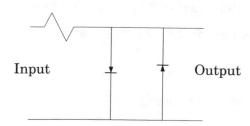

FIGURE 5.12: A limiting circuit.

correct by checking the sign on the current through the diode. If the sign is negative then our original assumption was incorrect and the diode is actually off. In this case we replace the diode with an open circuit and solve the circuit again. We could also use the opposite assumption, i.e., assuming the diode is off and then checking to see if the assumption was valid. Assuming the diode is on means

- assuming the voltage across the diode is equal to the threshold voltage (0.7 V in our case);

- checking the assumption by looking at the sign of the current through the diode. A positive current means that the assumption was correct and a negative sign on the current means that our assumption was incorrect.

Assuming the diode is off means

- replacing the diode with an open circuit;

- checking the assumption by calculating the voltage across the open circuit. If the voltage is less than the threshold voltage the assumption was correct and if the voltage is greater than the threshold voltage the assumption was incorrect.

We can see how this works through an example.

Example 5.4.1: Consider the circuit shown in Fig. 5.13, where we are asked to find the current I_o. Let's assume that the diode is on. This means that

$$V_{bc} = 0.7\,\text{V}$$

We have marked the nodes on the circuit diagram. Let's select node e to be the reference node. Then we proceed to each node:

Node a: $V_{ae} = 6\,\text{V}$

Node b: At node b we write the current law

$$I_{ab} = I_{be} + I_{bc} \qquad (5.2)$$

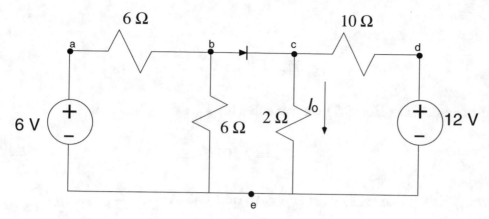

FIGURE 5.13: A circuit containing a diode.

Node c: Writing the current law at node c we obtain

$$I_{bc} = I_{ce} + I_{cd} \qquad (5.3)$$

Node d: $V_{de} = 12\,\mathrm{V}$

Writing the currents in terms of the voltages in the two equations found by writing the current law at nodes b and c we obtain

$$\frac{V_{ab}}{6} = \frac{V_{be}}{6} + I_{bc} \qquad (5.4)$$

$$I_{bc} = \frac{V_{ce}}{2} + \frac{V_{cd}}{10} \qquad (5.5)$$

Combining the two equations we get

$$\frac{V_{ab}}{6} = \frac{V_{be}}{6} + \frac{V_{ce}}{2} + \frac{V_{cd}}{10} \qquad (5.6)$$

Now we write the voltages in terms of the voltages at each node with respect to the reference node.

$$\frac{6 - V_{be}}{6} = \frac{V_{be}}{6} + \frac{V_{ce}}{2} + \frac{V_{ce} - 12}{10} \qquad (5.7)$$

We have one equation and two unknowns. Now assume that the diode is on.

$$V_{be} - V_{ce} = 0.7\,\mathrm{V}$$

or

$$V_{be} = V_{ce} + 0.7$$

Substituting this in Eq. (5.7) we can solve for V_{ce} as

$$V_{ce} = 2.107 \, \text{V}$$

Knowing V_{ce} we can easily calculate I_o to be 1.053 A.

Now we go back and check if our assumption was correct.

$$I_{ab} = I_{ce} + I_{cd} \tag{5.8}$$

$$= \frac{V_{ce}}{2} + \frac{V_{ce} - 12}{10} \tag{5.9}$$

$$= 1.053 - 0.989 \tag{5.10}$$

$$= 0.064 \tag{5.11}$$

which is positive. This validates our assumption.

Try repeating this example replacing the 10 Ω resistor with a 4 Ω resistor.

Assuming a diode is on means

- assuming the voltage across the diode is 0.7 V;

- checking the assumption by looking at the sign of the current. A positive sign indicates our assumption was true and a negative sign means our assumption was false.

Assuming the diode is off means

- replacing the diode with an open circuit;

- checking the assumption by calaculating the voltage across the open circuit. If the voltage is less than the threshold voltage the assumption is true. Otherwise the assumption is false.

In our analysis we admit only two possibilities: Either the diode is on or the diode is off.

What if we have more than one diode in the circuit? All that does is increase the number of possibilities. If we have two diodes in the circuit we will have four possibilities: both diodes on, both diodes off, first diode on and second diode off, and first diode off and second diode on. If we have n diodes in the circuit we will have 2^n possibilities. Solving such a circuit will be much more tedious; however, the technique remains the same: assume one possibility, solve the circuit, and check your assumption. If the assumption is not satisfied pick a different possibility.

5.5   SOLVING CIRCUITS USING LOAD LINES

In our solution of circuits containing diodes we have used an approximation to the diode I–V characteristic. That is, we assume that the diode is either on or off. In general, this is a reasonable approach. However, there may be situations where the approximation is not acceptable. In such a situation we can use a graphical approach.

A equation containing two unknowns defines a line. For example, the equation

$$6x - 2y = 10$$

corresponds to a line with a slope of 3 and a y intercept of -5. By this we mean that all points on such a line will satisfy this equation. Let's suppose we have another equation

$$2x + y = 10$$

All points on the line with a slope of -2 and a y intercept of 10 satisfy this equation. We have two (independent) equations and two unknowns, and hence we can find a unique solution to these equations. The solution has to satisfy both equations. Therefore, the point represented by the solution is a point on both lines.

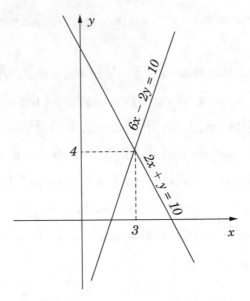

FIGURE 5.14: Graphical solution of two linear equations.

In other words, the solution is the point at the intersection of the two lines as shown in Fig. 5.14. This method of graphical solution is referred to as the load-line approach in electrical engineering.

In this particular case it would have been much simpler to have solved the equations algebraically rather than use the graphical method. However, when we start getting nonlinear terms in our equations, or when some information is only available in graphical form, the graphical solution method can be very handy.

Let's look at the following example.

Example 5.5.1: Consider the circuit shown in Figure 5.15. The relationship between the current through A and the voltage across A is given graphically in Figure 5.16. Following previous methods we would write the current law at node a.

$$I_{ba} = I_{ac}$$

From the circuit we can write an expression for I_{ba}

$$I_{ba} = \frac{10 - V}{10}$$

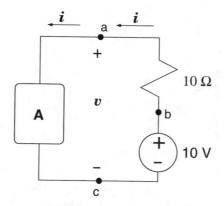

FIGURE 5.15: Circuit with nonlinear component.

If we could get an expression for I_{ac} in terms of V we could equate the two expressions and solve for V, which we can then use to find the various currents and voltages. Given that instead of an expression for I_{ba} we have a graph, how about expressing the relationship between I_{ba} and V graphically? Plotting this relationship we obtain the graph shown in Fig. 5.17. The graph shows the possible values for I_{ba} and V. The graph in Fig. 5.16 shows the possible pairs of values for I_{ac} and V when we consider the component A. What we need to do is find the value of V for which $I_{ba} = I_{ac}$. We can do that by finding the point at which the two graphs intersect. In order to do that we plot both of them on the same graph as shown in

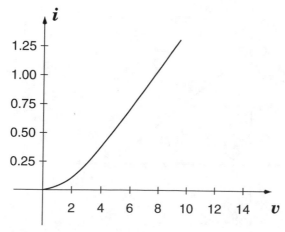

FIGURE 5.16: Nonlinear component rule.

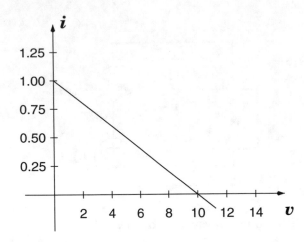

FIGURE 5.17: Load line.

Fig. 5.18. The point at which the two graphs intersect gives us the value of V and the values of I_{ba} and I_{ac}.

Let's repeat Example 5.3.1 using this graphical approach.

Example 5.5.2: Following the same reasoning as in Example 5.3.1 we obtain V_R as

$$V_R = \begin{cases} 0 & t < t_0 \\ ? & t_0 < t < t_1 \\ 0 & t > t_1 \end{cases}$$

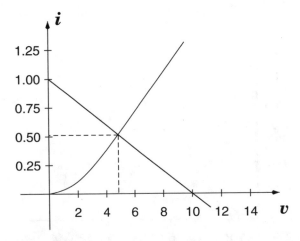

FIGURE 5.18: Graphical solution.

To find what happens between t_0 and t_1 this time let's use the load-line approach to find out what happens to V_D when $V(t)$ is greater than 0.7 V. We will use the ideal diode characteristics for expressing the relationship between I_{ab} and V_{ab} and use the current law at node b, which gives us

$$I_{ab} = I_{bc}$$

Now all we need to do is find the relationship between I_{bc} and V_{ab}. The current I_{bc} is given by Ohm's law as

$$I_{bc} = \frac{V_{bc}}{R}$$

In order to write V_{bc} in terms of V_{ab} we can use c as the reference node and we get

$$V_{ab} = V_{ac} - V_{bc}$$

or

$$V_{bc} = V_{ac} - V_{ab}$$

which gives us

$$I_{bc} = \frac{V_{ac} - V_{ab}}{R}$$

Substituting $V_{ac} = V(t)$ we obtain the equation

$$I_{bc} = -\frac{1}{R}V_{ab} + \frac{V(t)}{R}$$

Note that the "constant" term in this equation is a function of time and depends on the (unknown) value of R. But the value of R (as long as it is finite and nonzero) does not really matter. The only thing that matters about $V(t)$ is

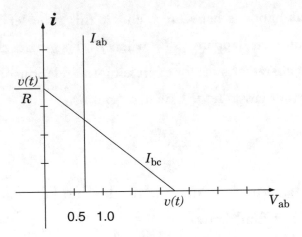

FIGURE 5.19: Graphical solution for ideal diode.

that it be greater than 0.7 V. We can see this from Fig. 5.19: as long as $V(t)$ is greater than 0.7 V the voltage across the diode will be 0.7 V. Therefore as $V_{bc} = V_{ac} - V_{ab}$,

$$
V_R = \begin{cases} 0 & t < t_0 \\ V(t) - 0.7 & t_0 < t < t_1 \\ 0 & t > t_1 \end{cases}
$$

Lets try to analyze a diode circuit with a less idealized diode characteristic.

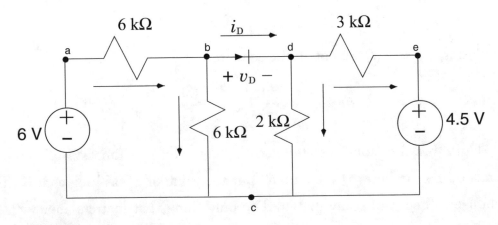

FIGURE 5.20: Diode circuit.

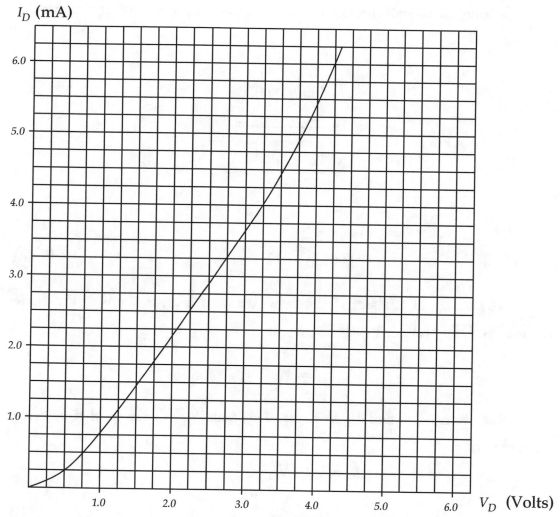

FIGURE 5.21: Diode component rule.

Example 5.5.3: Consider the circuit shown in Fig. 5.20. Suppose the diode has a characteristic shown in Fig. 5.21. We will find a relationship between the current through the diode I_D and the voltage across the diode V_D.

Let's write the current law at nodes b and d.

$$I_{ab} = I_{bc} + I_D \Rightarrow I_D = I_{ab} - I_{bc}$$
$$I_D = I_{dc} + I_{de}$$

Writing the current through the resistors in terms of the voltages across the resistors.

$$I_D = \frac{V_{ab}}{6000} - \frac{V_{bc}}{6000}$$

$$= \frac{V_{ac} - V_{bc}}{6000} - \frac{V_{bc}}{6000}$$

$$= \frac{12 - 2V_{bc}}{6000}$$

or

$$6000\, I_D = 12 - 2V_{bc} \tag{5.12}$$

We have I_D in terms of V_{bc} but what we want is I_D in terms of the diode voltage V_D. Using KVL we can write V_{bc} as

$$V_{bc} = V_D + V_{dc}$$

Now we need to find V_{dc} in terms of I_D and V_D. We can do that by using KCL at node d.

$$I_D = I_{dc} + I_{de}$$

$$= \frac{V_{dc}}{2000} + \frac{V_{de}}{3000}$$

$$= \frac{V_{dc}}{2000} + \frac{V_{dc} - V_{ec}}{3000}$$

$$= \frac{5V_{dc} - 9}{6000}$$

From which we can obtain an expression for V_{dc}

$$V_{dc} = \frac{6000\, I_D + 9}{5}$$

Substituting this in Eq. (5.12) we get

$$6000\, I_D = 12 - 2V_D - 2\left(\frac{6000\, I_D + 9}{5} \right)$$

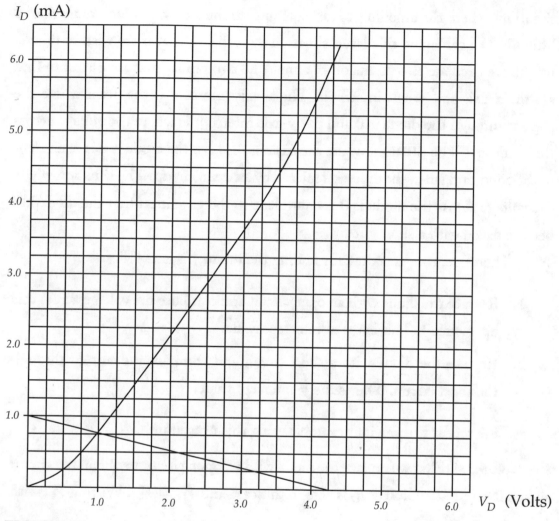

FIGURE 5.22:

Simplifying and writing I_D in units of milliamperes

$$I_D = 1 - \frac{5}{21}V_D$$

Drawing this line on the diode characteristics shown in Fig. 5.21, we obtain Fig. 5.22. We can see that the two curves intersect at $I_D = 0.8$ mA and $V_D = 1$ V.

In this example it took a little bit of manipulation to get the equation relating the current through the diode with the voltage across the diode. It is easy to see situations where the amount of manipulation required may be much more than a little bit. We can avoid this latter situation by realizing that if we wish to draw a line all we need are two points on the line. The two points that are easiest to find are the x-axis and y-axis intercepts. The x-axis intercept is the point where the current through the diode is 0 and the y-axis intercept is where the voltage across the diode is 0. The situation where the current through the diode is 0 is equivalent to the condition when the diode is replaced by an open circuit. The situation where the voltage across the diode is 0 is equivalent to the condition that the diode has been replaced with a short circuit.

Therefore, the procedure for finding the intercepts is as follows:

1. Replace the diode with an open circuit and compute the voltage across this open circuit. This is the x-intercept value.

2. Replace the diode with a short circuit and compute the current through the short circuit. This is the y-intercept value.

Let's repeat the previous example using this procedure.

Example 5.5.4: First we replace the diode with an open circuit to obtain the circuit shown in Fig. 5.23. Clearly I_D is zero in this case and $V_D = V_{bd}$. Writing V_{bd} with reference to node c

$$V_D = V_{bc} - V_{dc}$$

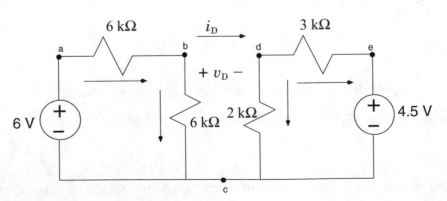

FIGURE 5.23: Circuit to find the x-axis intercept.

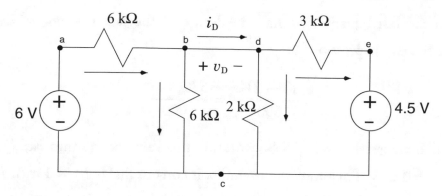

FIGURE 5.24: Circuit for finding the y-axis intercept.

Now writing KCL at nodes b and d we get $V_{bc} = 6$ V and $V_{dc} = 1.8$ V, which results in $V_D = 6 - 1.8 = 4.2$ V. So the first point on our line is $I_D = 0$, $V_D = 4.2$. To get the second point on the line we replace the diode with a short circuit. This gives us the circuit shown in Fig. 5.24. The voltage V_D is clearly 0 and I_D is the current flowing through the short circuit. To get this current we can write two KCL equations at nodes b and d and then use the fact of the short circuit. First, at node b:

$$
\begin{aligned}
I_D &= I_{ab} - I_{bc} \\
&= \frac{V_{ab}}{6000} - \frac{V_{bc}}{6000} \\
&= \frac{V_{ac} - V_{bc}}{6000} - \frac{V_{bc}}{6000} \\
&= \frac{12 - 2V_{bc}}{6000}
\end{aligned}
\tag{5.13}
$$

Then at node d:

$$
\begin{aligned}
I_D &= I_{dc} + I_{de} \\
&= \frac{V_{dc}}{2000} + \frac{V_{de}}{3000} \\
&= \frac{3V_{dc}}{6000} + \frac{2(V_{dc} - V_{ec})}{6000} \\
&= \frac{5V_{bc} - 9}{6000}
\end{aligned}
$$

where in the last equation we have made use of the fact that $V_{bc} = V_{dc}$. Now equating the two expressions for I_D.

$$\frac{12 - 2V_{bc}}{6000} = \frac{5V_{bc} - 9}{6000}$$

from which we get $V_{bc} = 21/7$. Substituting this value of V_{bc} into Eq. (5.13) we obtain $I_D = 1$ mA. Therefore, the second point on our line is $I_D = 1$ mA, $V_D = 0$. This is the same line that we obtained in the previous example.

Using the open circuit voltage and short circuit current to obtain two points, on a line rather than obtaining an equation for the line, is a systematic way of solving the same problem. In some cases it may be easily evident what the equation is for the line. However, the open circuit voltage/short circuit current approach is always available to us as an alternative.

5.6 SUMMARY

In this chapter we have introduced our first nonlinear component, the diode. We have shown a number of ways the diode is used. Circuits containing diodes can be treated in two ways depending on how we approximate the voltage–current relationship, or the component rule, for the diode. If we treat it as an on–off switch, we make an assumption as to the state it is in: on or off. Based on our assumption we solve the circuit and then check whether our assumption was correct. Whether the assumption turns out to be correct or incorrect at the end of this process, we know the state of the diode switch. The other approach to analyzing circuits containing diodes is used when the current–voltage relationship for the diode is given to us in the form of a graph. In such cases, we construct a load line whose intersection with the graph for the diode gives us the values of the voltage across the diode and the current through it. There are two ways of obtaining the load line: we can obtain the equation for the line or we can find two points on the line, which can be used to construct the line.

5.7   PROJECTS AND PROBLEMS

1. In the following circuit find V_R when

    · • $V_i = 1.8$ V.

    • $V_i = 0.4$ V.

    • $V_i = -1.8$ V.

    • $V_i = -0.4$ V.

    Assume an ideal diode with a threshold voltage of 0.7 V.

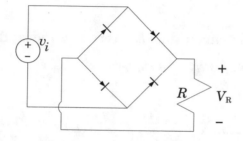

2. In the circuits shown below what is the value of V_{out} if (i) $V_{in} = 15$ V and (ii) $V_{in} = -5$ V? Assume $V_T = 0.7$ V.

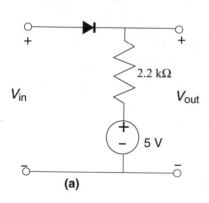

(a)

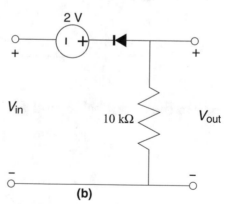

(b)

3. Find I_o in the circuit shown below.

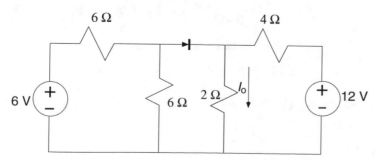

4. Find I_o in the circuit shown below.

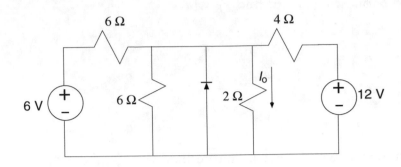

5. In the following circuit find I_o. Assume an ideal diode with a threshold voltage of 0.7 V. *Justify all assumptions.*

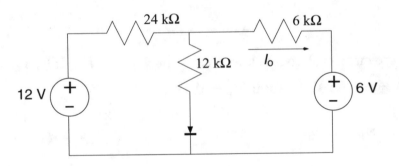

6. In the circuit below find V_o. Assume $V_T = 0.7$ V.

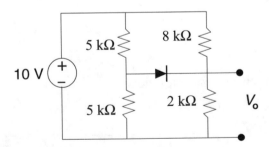

7. In the following circuit find V_o. Use the *I–V* characteristic for the diode shown.

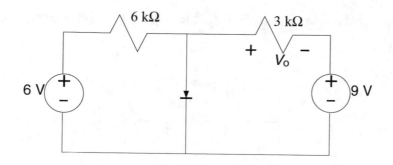

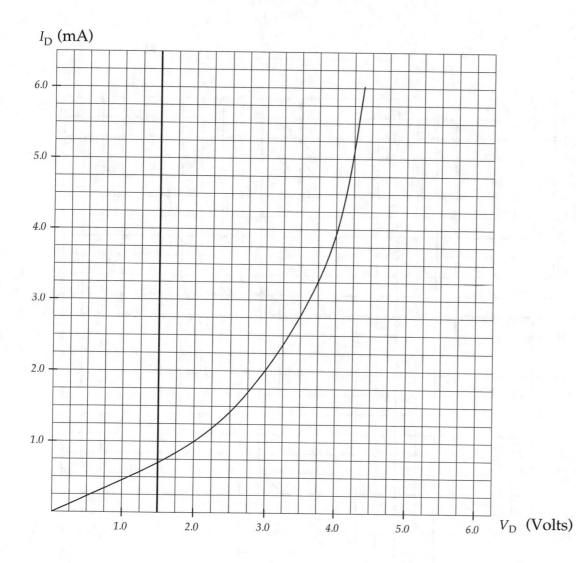

8. In the following circuit find I_o. Use the I–V characteristic for the diode shown.

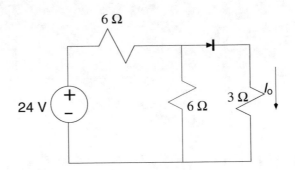

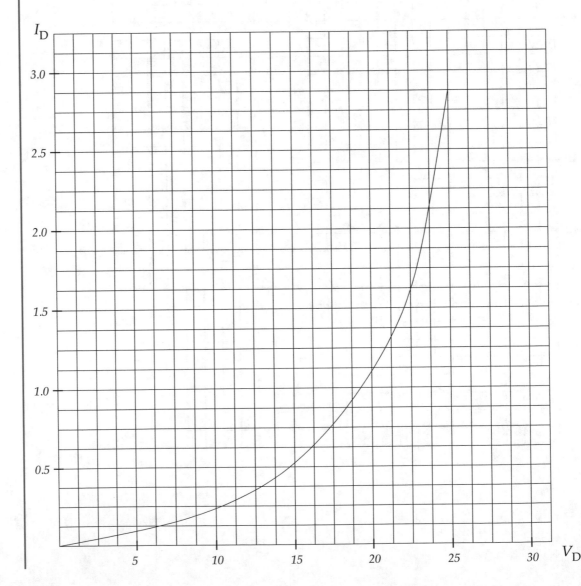

CHAPTER 6

Transistors

6.1 OVERVIEW

In this chapter we look at a component that revolutionized circuit design, the transistor. Understanding the transistor in all its glory is beyond the scope of this book; however, we can get some idea of its operation using the information we already have. We will analyze simple circuits with transistors and look at a few applications.

6.2 INTRODUCTION

The fact that the diode acts as a one-way valve permits us to do a number of interesting things. However, often what we need is not a valve but a tap that will control and amplify the flow of current. Modulating the flow of current through a component also lets us control the change of voltage across the component. We had earlier mentioned how a radio grabs sound transmissions from the air. The sound transmission is represented in the form of voltage fluctuations. However, these voltage fluctuations are so weak that we cannot use them to directly power a speaker. What we need is some method of amplifying these fluctuations.

When we turn on a tap, by expending a small amount of energy we can control the generation of a lot of force. The amount of force generated depends on the water pressure available, which is usually due to water being raised to a tank. The process of raising the water to the tank results in the storage of energy. Turning the tap does not require a lot of energy but it controls the large amount of energy stored in the tank. If the water pressure is high, a relatively frail individual can control a lot of power. In the case of the radio, the frail individual corresponds to the weak signal in the atmosphere, and what we need is for this weak signal to control the relatively large flow of current needed to drive a speaker. A component that permits this kind of *amplification* is the *transistor*. As in the water analogy, the transistor permits a weak signal to control the release of stored energy. The stored energy is in the form of an external voltage source. The transistor requires the external source to operate as a transistor. Thus, the transistor is called an *active* component, as opposed to resistors and capacitors, which are called *passive* components.

6.3   THE COMPONENT RULE

There are several different kinds of transistors, and you will learn about them in detail in more advanced courses. For now we will use one particular type of transistor called a bipolar junction transistor (*b jt*). There are two kinds of bipolar junction transistors. The kind we will look at is called an *npn* bipolar junction transistor and has the symbol shown in Fig. 6.1. Notice that the transistor has three terminals. They are called the *base*, the *collector*, and the *emitter*. The base of the transistor is where you apply the weak signal. The base–emitter junction acts like a diode. You need to increase the voltage V_{be} beyond the diode threshold before you can get any current to flow into the base. As in the previous chapter, we will assume that the diode threshold is 0.7 V. The current in the collector arm of the transistor is given by

$$I_c = \beta I_b$$

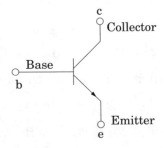

FIGURE 6.1: A bipolar junction transistor.

where β is a large number (typically about 100) *as long as the voltage difference V_{ce} is greater than a threshold.* This last point is very important because the base current I_b is only *controlling* the current I_c not generating it. The current is being generated by an external source and, if the the potential difference V_{ce} is not positive (or very small), this will block the flow of the current from the collector terminal to the emitter terminal. When we have a situation where the voltage V_{ce} is at its lowest value, we say that the transistor is *saturated.* In this case, the maximum value that I_c can take on is the current that will leave V_{ce} barely above the threshold. This threshold varies depending on the type of the transistor. A typical value is about 0.3 V. To make life easier for ourselves, we will approximate the threshold to be 0 V. Thus we will say the transistor is saturated if when I_c calculated using $I_c = \beta I_b$ would result in $V_{ce} \leq 0$. Therefore, the component rule for the *b jt npn* transistor is

$$I_b = 0 \quad \text{when} \quad V_{be} < 0.7 \tag{6.1}$$

$$I_c = \beta I_b \quad \text{when} \quad V_{ce} > 0 \tag{6.2}$$

The value of β depends on a number of different factors and can vary quite a bit. This can be a problem when designing circuits with specific characteristics. There are a number of ways of dealing with this problem, which you will see when you get a more detailed exposure to electronic circuits.

6.4 TRANSISTOR CIRCUITS

As in the case of circuits containing other components, our major tool for analyzing circuits containing transistors will be the Kirchhoff's laws, along with the rules relating the various currents in the transistor. Reiterating

1. The base current I_b will be zero if $V_{be} < 0.7$ V.

2. $I_c = \beta I_b$ whenever $V_{ce} > 0$.

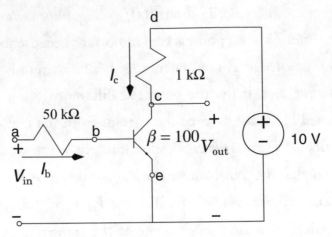

FIGURE 6.2: A common emitter amplifier.

If $I_b = 0$, this means the transistor is off. So, the first thing we do when analyzing a transistor circuit is calculate I_b. If I_b is greater than 0 we proceed to the next step, which is to calculate $I_c = \beta I_b$. We then check to see if V_{ce} is greater than 0. If it is, we are done with this part of the analysis. If not, we use the fact that $V_{ce} = 0$ in the rest of our analysis.

To see how we use these rules, consider the following example.

Example 6.4.1: A simple transistor amplifier circuit is shown in Fig. 6.2. We wish to calculate V_{out} for a given value of V_{in}. Let's suppose for the moment that V_{in} is 5 V. Can we calculate what the current into the base is going to be? From the figure we can see that the current into the base is the same as the current through the 50 kΩ resistor. We can use Ohm's law to calculate the current through the resistor.

$$I_{ab} = \frac{V_{ab}}{50 \times 10^3} = \frac{V_{ae} - V_{be}}{50 \times 10^3}$$

If there is any current flowing into the base, according to the component rules for the transistor, $V_{be} = 0.7$. The voltage V_{ae} is the same as V_{in}, therefore,

$$I_{ab} = \frac{5 - 0.7}{50 \times 10^3} = 86 \, \mu A$$

How did we know that there was current flowing into the base and therefore $V_{be} = 0.7$? We didn't. We simply made an assumption. If our assumption had been wrong, we would have obtained a negative value for I_{ab}.

Assuming the transistor is not saturated the collector current I_c is given by

$$I_c = \beta I_b = 100 \times 86 \times 10^{-6} \quad (6.3)$$
$$= 8.6 \text{ mA} \quad (6.4)$$

Now we must check if the transistor is saturated. If we have 8.6 mA flowing through the 1 kΩ resistor, the voltage across that resistor, V_{dc}, is 8.6 V. Then,

$$V_{out} = V_{ce} = V_{de} - V_{dc} = 10 - 8.6 = 1.4$$

V_{ce} is positive, the transistor is not saturated, and our calculation of I_c is valid.

What would happen if we changed the resistor from 50 to 10 kΩ? In this case the base current I_{ab} becomes

$$I_{ab} = \frac{5 - 0.7}{10 \times 10^3} = 0.43 \text{ mA}$$

If we now assume that the collector current is β times the base current with $\beta = 100$ we obtain the current into the collector as

$$I_{dc} = 100 \times 0.43 \text{ mA} = 43 \text{ mA}$$

But this would mean that the voltage across the 1 kΩ resistor V_{dc} would be 43 V. As V_{de} is 10 V this would mean V_{ce} would have to be -33 V. For V_{ce} to be negative the potential at the collector is lower than the potential at the emitter. This would be a violation of the requirements under which the collector current is β times the base current. Therefore, I_{dc} cannot be 43 mA.

In order to calculate the collector current, we use the fact that as the base current increases, the collector current increases until V_{ce} is 0. It cannot increase any further. This means that the maximum value for V_{dc} in this case is 10 V. For the voltage across a 1 kΩ resistor to be 10 V the current, by Ohm's law, I_c has to be 10 mA.

We should be careful about applying the $I_c = \beta I_b$ rule. It is valid as long as it does not violate the requirements on transistor operations.

Notice that even though the transistor looks considerably more complicated than the previous circuit components we have already looked at, analyzing a circuit containing transistors requires the same approach that we have used previously. In this last example we used Ohm's law to find the base current and the component rules for the transistor for the rest.

Let's look at a more complicated circuit.

Example 6.4.2 Consider the circuit shown in Fig. 6.3. We need to compute the output voltage V_{out}.

In order to use the component rules for transistors, we need to find the current I_b into the base of the transistor. In the previous example, the base current was the same as the current through a resistor; therefore, we used Ohm's law to find the current through the resistor and hence the base current. But now the situation is not as convenient. Therefore, we fall back on our tried and true approach: using the Kirchhoff's current law. Applying the current law to node b we get

$$I_{ab} = I_b + I_{bd} \tag{6.5}$$

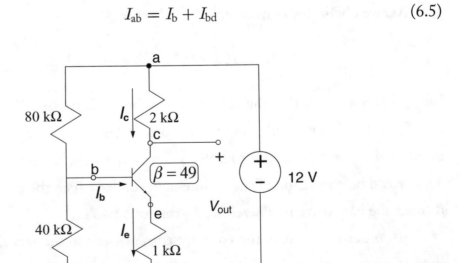

FIGURE 6.3: Another transistor amplifier.

or

$$I_b = I_{ab} - I_{bd} \tag{6.6}$$

Assuming d to be the reference node, we can write the two currents on the right in terms of the voltages using Ohm's law.

$$I_b = \frac{V_{ab}}{80\,\text{k}\Omega} - \frac{V_{bd}}{40\,\text{k}\Omega}$$

We can write these voltages in terms of the reference node

$$
\begin{aligned}
I_b &= \frac{V_{ad} - V_{bd}}{80\,\text{k}\Omega} - \frac{V_{bd}}{40\,\text{k}\Omega} \\
&= \frac{12 - V_{bd}}{80\,\text{k}\Omega} - \frac{V_{bd}}{40\,\text{k}\Omega}
\end{aligned} \tag{6.7}
$$

If we could now write I_b in terms of V_{bd} we would have an equation with a single unknown. Let's explore other ways of expressing I_b. Looking at Fig. 6.3 we see that we can write another expression including I_b using the KCL. Notice that I_c and I_b enter the transistor and I_e leaves the transistor. Therefore,

$$I_b + I_c = I_e$$

From the transistor component rules we know that if V_{ce} is greater than zero, $I_c = \beta I_b$. Therefore

$$I_e = (\beta + 1)I_b$$

(We will later check the assumption that V_{ce} is greater than 0.) I_e is the current through the 1 kΩ resistor, therefore

$$I_e = (\beta + 1)I_b = \frac{V_{ed}}{1\,\text{k}\Omega}$$

or

$$I_b = \frac{V_{ed}}{(\beta + 1) \times 1\,\text{k}\Omega}$$

We have an expression for I_b in terms of V_{ed}. What we wanted was an expression for I_b in terms of V_{bd}. Therefore, we need to write V_{ed} in terms of V_{bd}. Looking at Fig. 6.3 we can see that

$$V_{bd} = V_{be} + V_{ed}$$

If I_b is not zero, we know that V_{be} is 0.7 V. Therefore,

$$V_{ed} = V_{bd} - 0.7$$

and

$$I_b = \frac{V_{bd} - 0.7}{(\beta + 1) \times 1\,k\Omega}$$

as $\beta = 49$, this means that

$$I_b = \frac{V_{bd} - 0.7}{50 \times 1\,k\Omega} \tag{6.8}$$

Substituting this expression for I_b into Eq. (6.7) we obtain

$$\frac{V_{bd} - 0.7}{50 \times 1\,k\Omega} = \frac{12 - V_{bd}}{80\,k\Omega} - \frac{V_{bd}}{40\,k\Omega} \tag{6.9}$$

Solving for V_{bd} we obtain $V_{bd} = 2.85$ V. Substituting this value of V_{bd} in Eq. (6.8) we obtain $I_b = 0.043$ mA.

Our goal was to find V_{out}. From Fig. 6.3

$$V_{out} = V_{cd}$$

we know that V_{ad} is 12 V. We also know from the KVL that

$$V_{ad} = V_{ac} + V_{cd}$$

The voltage V_{ac} is the voltage across the 2 kΩ resistor, which is equal to $I_c \times$ 2 kΩ. Knowing that $I_c = \beta I_b$ we can calculate $I_c = 2.107$ mA and $V_{ac} = 4.214$ V. Therefore,

$$V_{out} = 12 - 4.214 = 7.786 \text{ V}$$

Before we can declare this to be our answer, we need to check the assumption that V_{ce} is greater than 0. From Fig. 6.3

$$V_{ad} = 12 = V_{ac} + V_{ce} + V_{ed}$$

Therefore,

$$V_{ce} = 12 - V_{ac} - V_{ed} = 12 - 4.214 - V_{ed}$$

As,

$$V_{ed} = V_{bd} - 0.7 = 2.85 - 0.7 = 2.15 \, V$$

V_{ce} is 5.636 volts, which is greater than 0. Hence our assumption, and the answer which relied on that assumption, is correct.

This example was more complicated than the one before it. However, notice that all we needed to analyze the circuit were the Kirchhoff's laws and the component rules. Our simple tools are again sufficient for the task.

Now that we have some idea about how to analyze simple transistor circuits, let's look at a few of the many transistor applications.

6.5   TRANSISTOR AMPLIFIERS

One of the most common uses of transistors are as amplifiers. In fact, the two circuits we analyzed in the previous section are both examples of amplifiers. The dictionary defines *amplify* as *to render larger, more extended, or more intense*. However, looking at Example 6.4.1 the results seem to contradict our naming of the circuit an amplifier. For an input of 5 V we got an output of 1.4 V; the magnitude of the output is less than the magnitude of the input. So, how can we say that the transistor is acting as an amplifier? The reason for this seeming contradiction is that when we say the circuit acts as an amplifier, we mean that it amplifies fluctuations, or changes, in the voltage at its input.

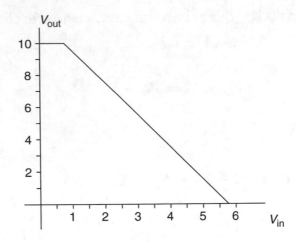

FIGURE 6.4:

For this circuit let's find V_{out} for different values of V_{in} and plot them as shown in Fig. 6.4. We can see from the figure, that as the input changes between 0.7 and 5.7 V, the output changes between 10 and 0 V. Thus, a change of 5 V at the input gets amplified to a change of 10 V at the output. This is a doubling of the input fluctuation.

There are a few little problems though. From the figure and from the computations, we can see that when the input voltage is low, the output voltage is high and vice versa. This is not really a problem because we are often more interested in the fluctuations of the voltages rather than the actual values. Also, there are simple ways of reversing this effect. A more serious problem is that this amplifier will only work when the input is positive. In fact, the transistor will only turn on when the input is greater than 0.7 V. This means that if the input looks like the waveform shown in Fig. 6.5(a), the output will look like the waveform shown in Fig. 6.5(b); all values of the input that are less than 0.7 V have been clipped.

Because most signals that we wish to amplify, such as speech signals, actually fluctuate between positive and negative values, this is a serious problem. The way we get around this problem is by *biasing* the transistor input. By biasing we mean that we add a fixed amount of voltage to the signal we want to amplify so that that the sum of the two voltages remains in the region in which we can get

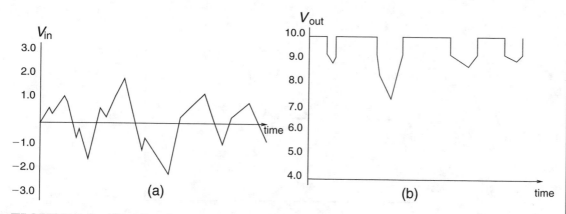

FIGURE 6.5: Clipping during amplification.

amplification. Consider the waveforms in Fig. 6.6. We start out with the same waveform as in Fig. 6.5(a). We then add a constant bias that makes the most negative point of the waveform rise to 0.7 V. This signal is then applied to the input of the transistor amplifier of Fig. 6.2. The bias is then removed from the amplified signal.

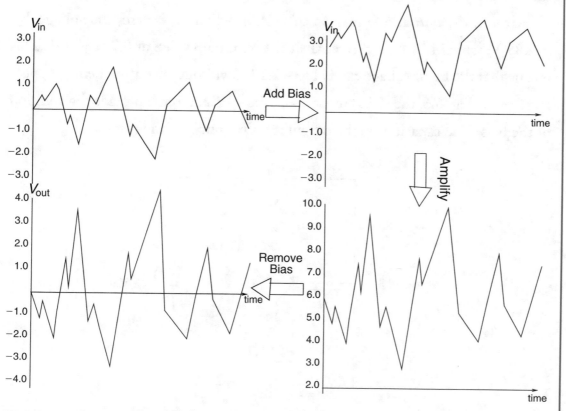

FIGURE 6.6: Biased amplifier.

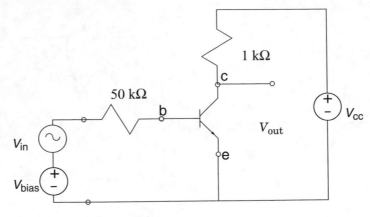

FIGURE 6.7: Transistor with amplifier.

We can represent the process of addition of bias as shown in Fig. 6.7.

Biasing a transistor is an important part of designing a transistor amplifier; we will leave the details to later courses. For now let's take look at an example of a biased transistor amplifier. Consider the circuit shown in Fig. 6.8. Notice that this is simply the circuit of Fig. 6.3 with an input terminal with a capacitor and a capacitor at the output terminal. In this circuit when V_{in} is zero, the voltage V_{bd} is 2.85 V, which is 2.15 V greater than the minimum value of 0.7 required. This means that the input voltage can fall to -2.15 V without the output being clipped. In terms of Fig. 6.6, the 80 kΩ resistor and the 40 kΩ resistors provide the addition of the bias. The capacitor on the output terminal removes the bias.

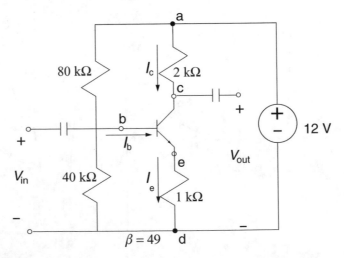

FIGURE 6.8: A biased amplifier.

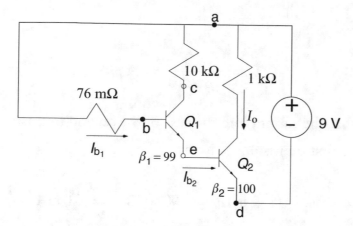

FIGURE 6.9: A current amplifier.

Finally, the transistor amplifier can be used as a current amplifier with some interesting (and simple to build) applications.

Example 6.5.1: Consider the circuit shown in Fig. 6.9. Let's calculate the current I_o. We begin by calculating the current into base of the first transistor (Q_1). This is the current through the $76\,\text{M}\Omega$ resistor. In order to obtain the current I_{ab}, we need the voltage V_{ab}. Picking node d as the reference node, we can write V_{ab} in terms of voltages with respect to the reference node.

$$V_{ab} = V_{ad} - V_{bd}$$

From the circuit we can see that $V_{ad} = 9$ V. Also, if the two transistors are on

$$V_{be} = V_{ed} = 0.7\,\text{V}$$

then

$$V_{bd} = V_{be} + V_{ed} = 1.4\,\text{V}.$$

Therefore,

$$V_{ab} = 9 - 1.4 = 7.6\,\text{V}$$

and

$$I_{ab} = \frac{V_{ab}}{76 \times 10^6} = .1\,\mu A$$

Noting that the current into the base of the second transistor is the emitter current from the first transistor, we have

$$I_{b_2} = (1 + \beta_1)I_{b_1} = 100 \times 10^{-7} = .01 \text{ mA}$$

Because the desired current I_o is the collector current for the second transistor

$$I_o = \beta_2 I_{b_2} = 100 \times 10^{-5} = 1 \text{ mA}$$

Notice that the current at the base of the first transistor has been amplified 10,000-fold.

You can use this particular circuit in a number of interesting ways. Replace the 1 kΩ resistor with an light emitting diode (LED) and then remove the 76 MΩ resistor and you have the touch key shown in Fig. 6.10. When your finger completes the circuit, a very minute amount of current will flow into the base of the first transistor. The amplification provided by the two transistors is sufficient to generate enough current to light up the LED. Try to build this circuit.

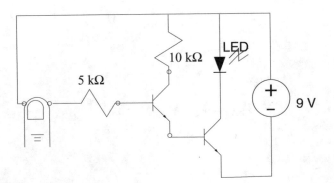

FIGURE 6.10: A touch-sensitive circuit.

6.6 TRANSISTOR LOGIC GATES

In the amplifier circuits we made use of the 'tap-like' characteristic of the transistor. Another characteristic of the transistor that has proven highly useful is its switch-like behavior. We can view the behavior of the circuit in Example 6.4.1 in a slightly different manner than we have done previously. For input voltages below 0.7 V, the transistor is in a state where the output has a large value. For input voltages greater than 5.7 V, the transistor is in a different state where the output voltage is zero. Thus, the transistor can be viewed as a voltage-controlled switch with two states. The behavior of the transistor for input voltages between 0.7 and 5.7 V can be viewed as a transition between the states.

This switch-like behavior can be used to build digital logic gates.

6.6.1 Inverter

One way you can use a transistor with some diodes is to implement an inverter or a NOT gate. The implementation is shown in Fig. 6.11. In order for the transistor to be turned on, we need to have 0.7 V across the base-emitter junction and 0.7 V across each of the diodes. Therefore, until the input voltage reaches 2.1 V, the transistor is turned off. If the transistor is turned off, there is no current through the 1 kΩ resistor and the voltage across the resistor is zero. Therefore, for V_{in} less than 2.1 V, V_{out} is 5 V. When the input voltage gets to be greater than 2.1 V, the

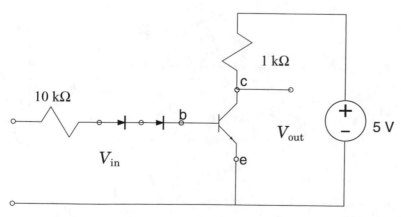

FIGURE 6.11: A NOT gate.

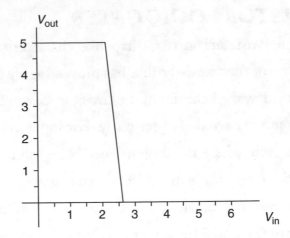

FIGURE 6.12: Input–output behavior of the NOT gate.

transistor turns on and V_{out} decreases. We can calculate that for V_{in} greater than 2.6 V V_{out} is 0. The behavior of this circuit is plotted in Fig. 6.12. Notice that the transition region is quite small. When the input is low (voltage less than 2.1 V), the output is high, and when the input is high (voltage greater than 2.6 V), the output is low. This is a description of a NOT gate.

6.6.2 NOR Gate

A NOR gate is a universal gate. That is, we can use NOR gates to construct all other digital gates. So, it is interesting to see how we can build a NOR gate using a transistor. A possible design for a two input NOR gate is shown in Fig. 6.13.

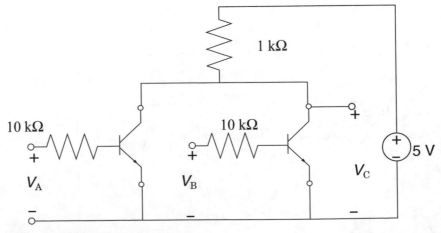

FIGURE 6.13: A two input NOR gate.

Assuming 5 V to be logic 1 and 0 V to be logic 0 we can build the truth table for this circuit where the inputs are V_A and V_B, and the output is V_C. If both inputs are low, which in the case of this circuit means that the inputs have a value less than 0.7 volts both transistors will be off. If the transistors are off no current will flow through the 1 kΩ resistor. Using KVL the output V_C is equal to the 5 V minus the drop across the 1 kΩ resistor. Therefore, $V_C = 5$ V. If either of the transistors is on there is a path for the current to flow through the 1 kΩ resistor. For $\beta = 100$ if either V_A or V_B is greater than 1.2 V that particular transistor enters saturation mode and $V_C = 0$ V. The truth table for this circuit is

V_A	V_B	V_C
0	0	5
5	0	0
0	5	0
5	5	0

This is clearly the truth table for a NOR gate. The circuit provides some wiggle room as well. The low value does not have to be exactly zero and the high value can vary considerably from 5 V.

By combining diodes and transistors, or multiple transistors in different ways, we can build all the different logic gates used to build digital circuits.

6.7   TRANSISTOR SWITCHES

The transistor allows us to use very low current values to control much larger currents. Two such switches are shown in Figs. 6.14 and 6.15. In each case the physical switch switches the base current, which is very low. Therefore, the switch does not suffer from as much wear and tear as it would have to if the switch was trying to handle the current through the equipment. The switch in Fig. 6.14 is a normally-off switch, that is, there is no current flowing through the equipment

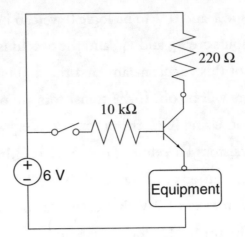

FIGURE 6.14: A normally off switch.

when the switch is off. It resembles several of the other transistor circuits we have looked at.

The switch in Fig. 6.15 is a bit different. This is a normally-on switch. Before the switch is closed the transistor is off and there is no current flowing into the collector. Instead all the current flows through the equipment. When the switch is closed the transistor enters saturation and all the current flows into the collector bypassing the equipment, thus turning off the flow of current to the equipment.

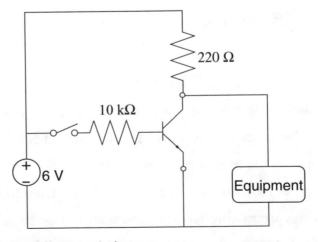

FIGURE 6.15: A normally on switch.

6.8 SUMMARY

In this chapter we introduced the transistor, the most versatile and widely used component in electronic circuits. We introduced the component rule for the bipolar junction transistor and examined several applications of transistors.

6.9 PROBLEMS

1. In the circuit below find I_c and V_{out} for (a) $V_{in} = 4\,V$ and (b) $V_{in} = 10\,V$. Assume $\beta = 100$.

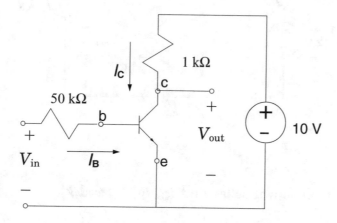

2. For the circuit shown below find I_c and V_{out} for $V_{in} = 2\,V,\ 3\,V,\ 4\,V,\ 5\,V,$ $6\,V,\ 7\,V,\ 8\,V,\ 9\,V,$ and $10\,V$. Present your answer in two ways: (a) Make a table with columns for V_{in}, I_c, and V_{out}. (b) Plot V_{out} and I_c versus V_{in}.

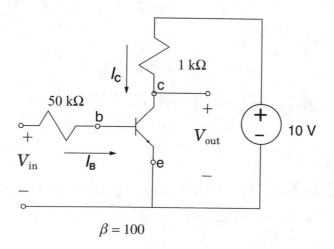

$\beta = 100$

3. In the following circuit find I_c and V_{out}. You can assume that the transistor is not saturated (in other words you don't have to show that it is not saturated).

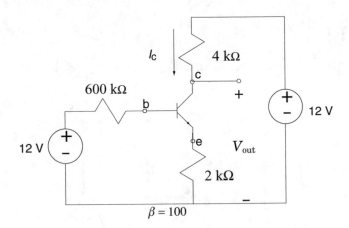

4. In the circuit shown below find I_C, I_B, I_E, and V_{out}.

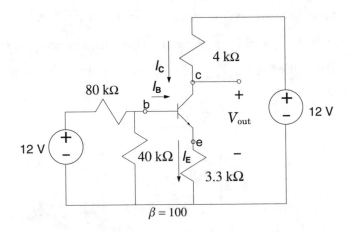

5. In the following circuit find R_1, R_2, R_E, and R_C such that $I_E = 1\,\text{mA}$, $V_{ed} = 3.3\,\text{V}$, $V_{bd} = 4\,\text{V}$, and $V_{cd} = 8\,\text{V}$.

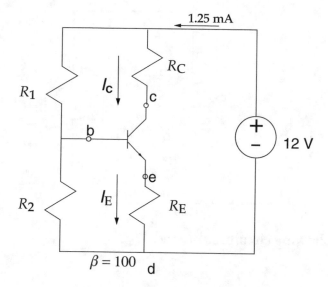

6. For the circuit shown below find V_{out} for $V_{in} = 1.6$ V, 1.8 V, 2.0 V, 2.2 V, 2.4 V, 2.6 V, 2.8 V, and 3 V. Present your answer in two ways: (a) Make a table with columns for V_{in} and V_{out}. (b) Plot V_{out} versus V_{in}.

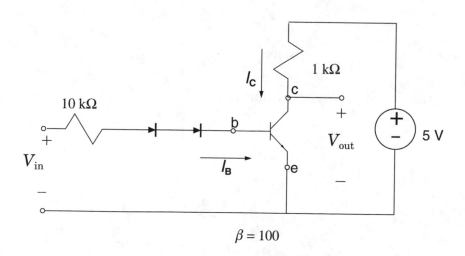

7. In the following circuit the source A generates a current of $10\,\mu$A in the presence of light. Pick resistors R_c and R_e so that when light shines on A, the voltage V_{out} is 5 V and V_{ce} is 1 V.

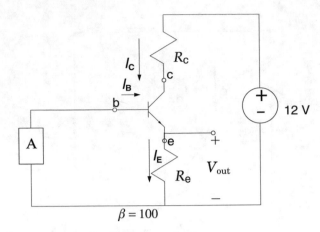

$\beta = 100$

8.   In the following circuit find I_c and V_{out}.

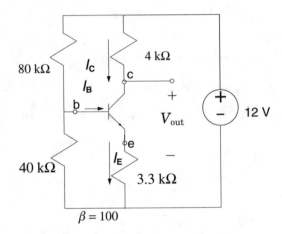

$\beta = 100$

9.   In the circuit shown below find I_c and V_{ce}.

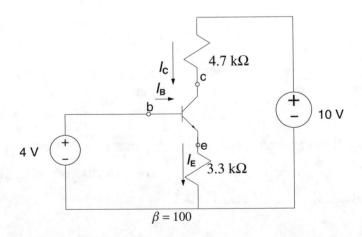

$\beta = 100$

10. In the following circuit find V_{out}.

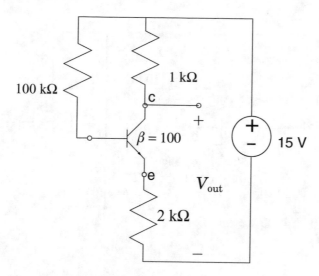

11. In the circuit below find I_c and V_{ce}.

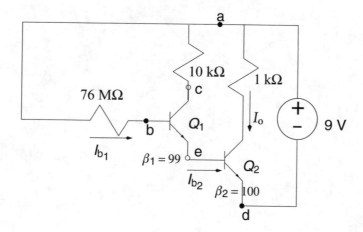

12. In the following circuit find I_c and V_{out}, if $V_{in} = 3\,\text{V}$.

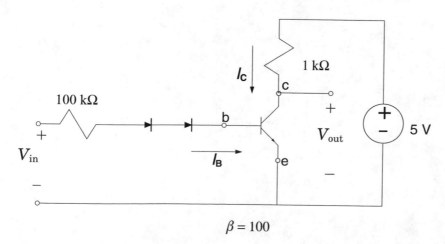

13.   In the circuit shown below find I_o.

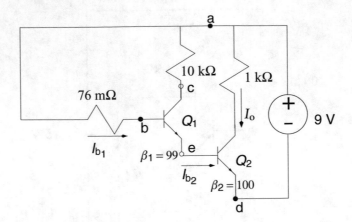

C H A P T E R 7

Operational Amplifiers

7.1 OVERVIEW

In this chapter we will learn about operational amplifiers and a number of circuits that we can construct using operational amplifiers.

7.2 INTRODUCTION

In this chapter we will look at a wonderful "component" called the operational amplifier or *op-amp* for short. The word component is in quotes because the op-amp is itself quite a complicated circuit containing resistors, capacitors, and transistors. However, we are not concerned with the internal construction of the op-amp but rather how the output of the op-amp is related to the input of the op-amp. In other words we want a component rule for the op-amp, which we can use in our analysis of circuits containing op-amps. In this context we treat the op-amp as a single component with some surprisingly simple rules that can be used to relate the voltage at the output terminals of the op-amp to the voltage at the input terminals. We do this by using a simple model of the op-amp that is reasonably accurate in terms of its functional behavior.

7.3 COMPONENT RULE

There are actually several models of the op-amp that are used in practice (one of them is shown in Fig. 7.2). The symbol for the op-amp is shown in Fig. 7.1. Notice in Fig. 7.1 that the op-amp is connected to a power supply. Like the transistor

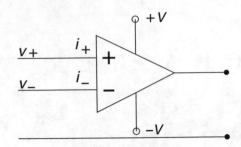

FIGURE 7.1: A symbol for an operational amplifier.

an op-amp requires an external power source to function as an op-amp. This makes it an *active* component like the transistor. Because we know that an op-amp cannot function without an outside power source we often do not include the power supply connection in the circuit diagrams.

In the model for the op-amp shown in Fig. 7.2 we have ignored the external power supply connections. In order to calculate the voltage V_{out} at the output of the op-amp, lets label the circuit as shown in Fig. 7.2. By the Kirchhoff's voltage law

$$V_{out} = V_{ac} = V_{ab} + V_{bc}$$

As there is no current flowing through the resistor the voltage across the resistor $V_{ab} = 0$. Therefore,

$$V_{out} = V_{ab} + V_{bc} = V_{bc}$$

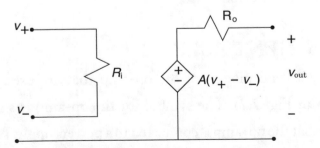

FIGURE 7.2: A model for an operational amplifier.

The voltage V_{bc} is the voltage generated by the funny looking voltage source. Thus,

$$V_{out} = V_{bc} = A(V_+ - V_-)$$

Or in other words, the output of the op-amp is a gain factor A multiplied by the difference in voltage $V_+ - V_-$. The value of the gain or amplification factor A is generally assumed to be very large (in fact we often assume that the gain of the op-amp approaches infinity). According to the model this means that for a small voltage $V_+ - V_-$ we can get a very large output voltage. However, you never get something for nothing and, in practice, the maximum output voltage we can have is limited by the voltage of the power supply V. The only way we can balance these two requirements of A being very large, and V_{out} being limited is to require that $V_+ - V_-$ be very small. In fact this difference is small enough that we can assume that $V_+ = V_-$. This assumption is called the *virtual short* assumption. Given the model of Fig. 7.2 if this true it necessarily implies that $I_+ = I_- = 0$. Otherwise $V_+ - V_-$ would have a nonzero value of $I_+ R_i$. These two operating assumptions

1. $I_- = I_+ = 0$
2. $V_- = V_+$

make up the component rule for the op-amp. From here on we will ignore the model and simply use the component rule.

As with the other components we can use the component rule along with the two Kirchhoff's laws to analyze op-amp circuits. Op-amp circuits rely on feedback for their operation. The feedback is provided by connecting a circuit component between the output of the op-amp and the terminal marked with a negative sign, called the *inverting terminal*. In our analysis it will not seem to make a difference whether the feedback connection is to the inverting terminal or the terminal marked with a positive sign (called the *non-inverting terminal*). However, in practice it makes a huge difference for reasons we will not go into here. Therefore, we will make a point of indicating that the feedback connection is to the inverting terminal.

Because the component rules for the op-amp are so simple, the appearance of an op-amp in a circuit generally makes it easier to analyze the circuit. While there may be differences in particular cases, in general the following procedure for analyzing op-amp circuits works most of the time:

1. Write the Kirchhoff's current law at the inverting terminal. Make use of the fact that $I_- = 0$.

2. Make use of the virtual short to find the voltage at the inverting terminal with respect to the reference node.

These instructions are a bit vague. We will make them more precise as we work through examples. One pitfall you should avoid is writing the current law at the output of the op-amp. This not because the Kirchhoff's laws do not hold at the op-amp output; regardless of the component, the Kirchhoff's laws always hold. It is because we have specified no rule for relating the output current of the op-amp to any other parameter in the circuit. The only way we can figure out the output current is to infer it from the rest of the circuit.

The number of useful circuits that can be built using op-amps, together with the other components we have looked at, are enormous. In what follows we look at some of the more popular configurations.

7.4   VOLTAGE FOLLOWER

The first circuit we look at is the simplest. As shown in Fig. 7.3, we connect the output of the op-amp (terminal c) directly to the inverting terminal of the op-amp.

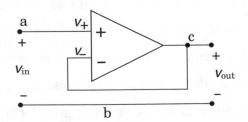

FIGURE 7.3: Voltage follower circuit.

Terminal c and terminal b make up the output terminals, while terminal a and terminal b make up the input terminals. Let us find the voltage between terminals c and b, or V_{out} in terms of V_{in}, which is the voltage between terminals a and b. The voltage at the noninverting terminal of the op-amp with respect to node b V_{ab} is simply V_{in}. Because the inverting terminal of the op-amp is connected directly to node c the voltage at node c with respect to node b, or V_{out} is the voltage at the inverting terminal with respect to node b. But by the virtual short assumption the voltage at the inverting terminal with respect to node b is the same as the voltage at the noninverting terminal with respect to node b. Therefore

$$V_{out} = V_{in}$$

This seems rather anticlimactic. Why not forget the op-amp and just connect node a to node c? To see the advantage provided by the op-amp let us return to the problem of the voltage divider we had looked at earlier. We wanted to obtain 5 V from a 9-V battery. One of the circuits we used is shown in Fig. 7.4. At the output terminal this circuit indeed does give us 5 V.

Suppose the circuit that requires the 5 V has a resistance of 100 kΩ. Let's model this circuit as a 100 kΩ load resistor and connect it to our 5-V "source" as shown in Fig. 7.5. Now let's calculate the voltage across the 100 kΩ load resistor. Picking b as our reference node we can write the current law at node a. Writing each current in terms of the node voltages we obtain

$$\frac{9 - V_{ab}}{40 \times 10^3} = \frac{V_{ab}}{50 \times 10^3} + \frac{V_{ab}}{100 \times 10^3}$$

FIGURE 7.4: Voltage divider circuit.

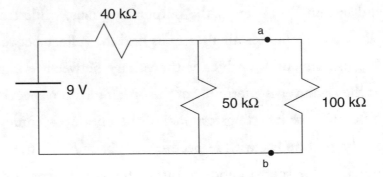

FIGURE 7.5: Voltage divider circuit with a 100 kΩ load resistor.

Solving for V_{ab} we obtain

$$V_{ab} = \frac{45}{11} = 4.09 \, \text{V}$$

This represents a drop of almost 20% from the required voltage level of 5 V. Instead of delivering 5 V to the load, we are delivering only about 4 V!

Now let's try something different. Connect the voltage divider to the voltage follower circuit and connect the load to the output of the voltage follower as shown in Fig. 7.6. We have again represented the load with a 100 kΩ resistor. Let us find the voltage across the 100 kΩ resistor.

From Fig. 7.6 we can see, by the virtual short assumption, that V_{out}, the voltage across the 100 kΩ resistor, is equal to the voltage at node a with respect to node b. Let us write the current law at node a using node b as the reference node.

$$\frac{9 - V_{ab}}{40 \times 10^3} = \frac{V_{ab}}{50 \times 10^3} + I_+$$

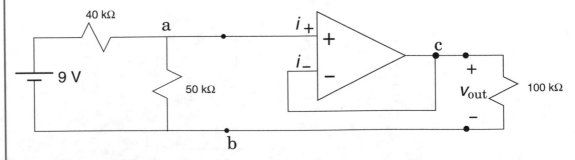

FIGURE 7.6: The voltage divider revisited.

But $I_+ = 0$ from assumption 1, therefore we have

$$\frac{9 - V_{ab}}{40 \times 10^3} = \frac{V_{ab}}{50 \times 10^3}$$

or

$$V_{ab} = 5\,\text{V}$$

Thus the voltage across the load is 5 V as desired. Unlike the case without the op-amp, connecting the load had no effect on the voltage across the 50 kΩ resistor.

How did this happen? If we trace back the steps in the analysis above we can see that the critical factor was the fact that the op-amp does not draw any current from the input stage. This means that we can add different combinations of components to the output stage without "loading" the input. Without drawing any current from the input, the op-amp voltage follower maintains the same voltage at the output that it sees at its input. The voltage follower thus *buffers* what is connected at its input from what is connected at its output terminals. This is useful in a large number of applications, such as when designing sensors. We do not want the sensing operation to disturb the operation of the system, so it is useful to have a buffer between the system and the sensor that protects the system but still allows the sensor to get on with its job. This is especially true when the system in question is a human being. When we connect electrodes to people to measure electrical activity in their brain or muscles we do not want any malfunction in the recording equipment to result in an attempt to draw current from the body. A voltage follower is a very useful tool for providing the isolation necessary.

While this is a very useful application of the op-amp, it is not the only one. In the next sections, we show some more applications of this component.

7.5   AMPLIFIER

As the name implies, one of the most popular application of the op-amp is as an amplifier. Amplification is obtained using feedback from the output. There are several different ways we can connect up the op-amp to obtain amplification.

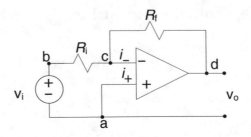

FIGURE 7.7: A simple op-amp circuit.

Two standard amplifier configurations are the inverting configuration and the non-inverting configuration.

7.5.1   Inverting Op-Amp

The inverting op-amp configuration is shown in Fig. 7.7. Suppose we wish to find V_o in terms of V_i given the resistance values R_i and R_f. Let's take node a to be our reference node and write the current law at node c

$$I_{dc} + I_{bc} = I_-$$

The currents I_{dc} and I_{bc} are currents through resistors. We can use Ohm's law to write these currents in terms of the voltages across these resistors.

$$\frac{V_{dc}}{R_f} + \frac{V_{bc}}{R_i} = I_-$$

Writing these voltages in terms of the node voltages and noting that $V_{ba} = V_i$ and $V_{da} = V_o$,

$$\frac{V_o - V_{ca}}{R_f} + \frac{V_i - V_{ca}}{R_i} = I_-$$

From our first operating assumption we have $I_- = 0$, and from the virtual short assumption we have $V_{ca} = V_{aa} = 0$. So our equation becomes

$$\frac{V_o}{R_f} + \frac{V_i}{R_i} = 0$$

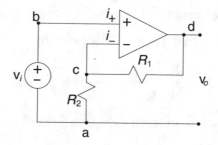

FIGURE 7.8: A noninverting amplifier circuit.

or

$$V_o = -\frac{R_f}{R_i} V_i$$

By suitably picking the values of R_f and R_i we can get an output voltage that is larger than the input voltage. Lest visions of an infinite power source begin dancing in your head,[1] remember that the output voltage is limited by the power supply which is being used to power the op-amp. This kind of amplifier is called an inverting amplifier. The "inverting" in the name comes from the fact that the output is negative of the input.

7.5.2 Non-Inverting Op-Amp

We can also construct a noninverting amplifier as shown in Fig. 7.8.

If we write the current law at node c we will obtain

$$I_{dc} = I_{ca} + I_-$$

The currents I_{dc} and I_{ca} are currents through resistors. Using Ohm's law we can write these currents in terms of the voltages across these resistors.

$$\frac{V_{dc}}{R_1} = \frac{V_{ca}}{R_2} + I_-$$

---

[1] Take the voltage fluctuations from your brain. Amplify them using an op-amp and run your television with it—an exercise in killing your brain with brain power!

Assuming a to be the reference node, writing these voltages in terms of node voltages, and using the fact that $V_{da} = V_o$, we obtain

$$\frac{V_o - V_{ca}}{R_1} = \frac{\dot{V}_{ca}}{R_2} + I_-$$

From assumption 1, $I_- = 0$, and from the virtual short assumption $V_{ca} = V_i$. Substituting these values into the equation, we get

$$\frac{V_o - V_i}{R_1} = \frac{V_i}{R_2}$$

or,

$$V_o = \left(1 + \frac{R_1}{R_2}\right)V_i$$

By picking R_1 to be much larger than R_2 we can get significant amplification, limited only be the power source. Notice that the polarity of the output voltage is the same as the polarity of the input voltage.

7.6   THE ANALOG COMPUTER

Long before PC came to symbolize personal computer, there were computers much much faster than any digital computer yet designed. These were analog computers that performed arithmetic and calculus. The building blocks for these analog computers were op-amps. While the analog computer is no more the circuits that performed the computations in the analog computer are still widely used. In this section we look at some of these circuits.

7.6.1 Adding Circuit

Consider the circuit shown in Fig. 7.9. This is a addition circuit.

$$V_0 = -\frac{R_0}{R_1}V_1 - \frac{R_0}{R_2}V_2 - \frac{R_0}{R_3}V_3$$

In order to analyze this, as before, we write the Kirchhoff's current law at the inverting terminal, which in the figure is node e.

$$I_{be} + I_{ce} + I_{de} + I_{fe} = I_-$$

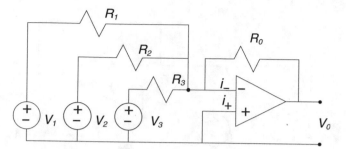

FIGURE 7.9: A summing amplifier circuit.

Writing the currents through the resistors in terms of the voltages across the resistors and noting that $I_- = 0$ we obtain

$$\frac{V_{be}}{R_1} + \frac{V_{ce}}{R_2} + \frac{V_{de}}{R_3} + \frac{V_{fe}}{R_0} = 0$$

Using node a as the reference node and writing the voltages in terms of the node voltages we get

$$\frac{V_{ba} - V_{ea}}{R_1} + \frac{V_{ca} - V_{ea}}{R_2} + \frac{V_{da} - V_{ea}}{R_3} + \frac{V_{fa} - V_{ea}}{R_0} = 0$$

Noting that $V_{ba} = V_1$, $V_{ca} = V_2$, $V_{da} = V_3$, and $V_{fa} = V_0$ we get

$$\frac{V_1 - V_{ea}}{R_1} + \frac{V_2 - V_{ea}}{R_2} + \frac{V_3 - V_{ea}}{R_3} + \frac{V_0 - V_{ea}}{R_0} = 0$$

Now we use the virtual short assumption to note that $V_{ea} = V_{aa} = 0$, and therefore,

$$\frac{V_1}{R_1} + \frac{V_2}{R_2} + \frac{V_3}{R_3} + \frac{V_0}{R_0} = 0$$

Solving for V_0 we obtain

$$V_0 = -\frac{R_0}{R_1}V_1 - \frac{R_0}{R_2}V_2 - \frac{R_0}{R_3}V_3$$

If we pick all the resistors to be identical we get

$$V_0 = -(V_1 + V_2 + V_3)$$

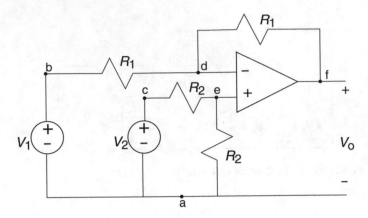

FIGURE 7.10: A difference circuit.

which is the negative of the sum of the inputs. We can easily get rid of the negative sign by using an inverting op-amp with unity gain. By letting the resistors take on different values we can design a circuit that will give us a weighted sum of the inputs. However, all the weights have to be the same sign. In the next application we see how we can change the sign.

7.6.2   Subtraction Circuit

Consider the circuit shown in Fig. 7.10. Notice that unlike the previous circuits there are components connected to the noninverting terminal of the op-amp. Despite this change the analysis proceeds as it did for the previous circuits. We first write the KCL at the inverting terminal, which in this case is node d.

$$I_{bd} + I_{fd} = I_-$$

Writing the currents through the resistors in terms of the voltages across the resistors and noting that $I_- = 0$ we get

$$\frac{V_{bd}}{R_1} + \frac{V_{fd}}{R_1} = 0$$

Multiplying both sides by R_1

$$V_{bd} + V_{fd} = 0$$

Picking node a as the reference node we can write the voltages in terms of the node voltages and use the fact that $V_{ba} = V_1$ and $V_{fa} = V_o$ to obtain

$$V_1 - V_{da} + V_o - V_{da} = 0$$

or,

$$V_o = 2V_{da} - V_1 \tag{7.1}$$

At this point in the previous analyses, we used the virtual short assumption to set the voltage at the inverting terminal equal to the voltage at the noninverting terminal. We do the same here and set $V_{da} = V_{ea}$. However, unlike earlier analyses where the noninverting terminal was directly connected to the reference node, we have to do some work to get the value of V_{ea}. Fortunately, it is not a lot of work. Writing the current law at node e we get

$$I_{ce} = I_{ea} + I_{+}$$

Writing the currents through the resistors in terms of the voltages across the resistors and using the fact that $I_{+} = 0$ we get

$$\frac{V_{ce}}{R_2} = \frac{V_{ea}}{R_2}$$

Multiplying both sides by R_2, writing the voltages in terms of node voltages, and substituting $V_{ca} = V_2$ we get

$$V_2 - V_{ea} = V_{ea}$$

or

$$V_{ea} = \frac{1}{2}V_2$$

Substituting this value of V_{ea} for V_{da} in Eq. (7.1) we get

$$V_o = V_2 - V_1$$

Notice that in the circuit of Fig. 7.10 we have four resistors but only two distinct values. If we had allowed all four resistor values to vary we would have obtained a relationship of the form

$$V_o = \alpha V_2 - \beta V_1$$

where the values of α and β would depend on the values of the four resistors.

7.6.3   Implementing Algebraic Equations

By combining the strategies used in the addition and subtraction circuits we can implement any algebraic equation using op-amps. Consider the following example:

Example 7.6.1: We are asked to implement the equation

$$V_o = 2V_1 - 3V_2 - 2V_3 \tag{7.2}$$

using op-amps. Let's first try to implement the equation directly. Sometimes it might be easier to implement the negative of the equation and then use an inverting op-amp to fix the polarity. Notice that V_2 and V_3 have the same polarity, which is opposite to that of V_1, so if we were to implement this using a single op-amp the V_2 and V_3 inputs would go to one terminal while the V_1 input would go to the other terminal. Based on the polarities the V_2 and V_3 inputs would go to the inverting terminal, while the V_1 input would go to the noninverting terminal. Therefore, the circuit would look as shown in Fig. 7.11.

To find the values of the various resistors, let's write the current law at the noninverting input, or node e.

$$I_{de} + I_{ce} + I_{ge} = I_-$$

Writing the currents through the resistors in terms of the voltages across the resistors and using the fact that $I_- = 0$

$$\frac{V_{de}}{R_3} + \frac{V_{ce}}{R_2} + \frac{V_{ge}}{R_o} = 0$$

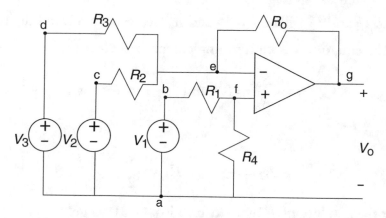

FIGURE 7.11: Implementing an algebraic equation.

Picking node a to be the reference node and writing the voltages in terms of the node voltages,

$$\frac{V_{da} - V_{ea}}{R_3} + \frac{V_{ca} - V_{ea}}{R_2} + \frac{V_{ga} - V_{ea}}{R_o} = 0$$

Now substitute the values of V_{da}, V_{ca}, and V_{ga}

$$\frac{V_3 - V_{ea}}{R_3} + \frac{V_2 - V_{ea}}{R_2} + \frac{V_o - V_{ea}}{R_o} = 0$$

Expanding this out

$$R_o R_2 V_3 - R_o R_2 V_{ea} + R_o R_3 V_2 - R_o R_3 V_{ea} + R_2 R_3 V_o - R_2 R_3 V_{ea} = 0$$

or

$$R_2 R_3 V_o = (R_o R_2 + R_o R_3 + R_2 R_3) V_{ea} - R_o R_3 V_2 - R_o R_2 V_3$$

Dividing both sides by $R_2 R_3$ we obtain

$$V_o = \frac{R_o R_2 + R_o R_3 + R_2 R_3}{R_2 R_3} V_{ea} - \frac{R_o}{R_2} V_2 - \frac{R_o}{R_3} V_3 \qquad (7.3)$$

Looking at the circuit and using the fact that we will find V_{ea} by using the virtual short assumption it is clear that V_{ea} will be in terms of V_1. Therefore, the coefficients of V_2 and V_3 will be R_o/R_2 and R_o/R_3 respectively in the final equation. Looking at the equation we want to implement, it is clear that

$$\frac{R_o}{R_2} = 3 \quad \text{and} \quad \frac{R_o}{R_3} = 2$$

or

$$R_2 = \frac{R_o}{3} \qquad R_3 = \frac{R_o}{2} \tag{7.4}$$

Substituting these values of R_2 and R_3 into Eq. (7.3) we get

$$V_o = 6V_{ea} - 3V_2 - 2V_3$$

Using the virtual short assumption and writing the current law at node f we obtain

$$V_{ea} = V_{fa} = \frac{R_4}{R_1 + R_4} V_1$$

From Eq. (7.2) we know that

$$6V_{ea} = 6\frac{R_4}{R_1 + R_4} V_1 = 2V_1$$

which will be true if

$$R_1 = 2R_4 \tag{7.5}$$

Now all we need to do is to pick resistor values that satisfy Eqs. (7.4) and (7.5). In practice not all resistor values may be available and there may be other constraints as well. For this example we do not worry about those constraints so we have an infinite number of possible solutions. One solution would be to pick $R_o = 6\,\text{k}\Omega$. From Eq. (7.4) this would mean $R_2 = 2\,\text{k}\Omega$ and $R_3 = 3\text{k}\Omega$. If we pick $R_4 = 1\,\text{k}\Omega$, from Eq. (7.5), $R_1 = 2\,\text{k}\Omega$.

7.6.4 Integration

In order to perform integration using op-amps we have to broaden our horizons a bit and use capacitors as well as resistors in our circuit. An integrator circuit is shown in Fig. 7.12. To see that this circuit really does perform integration, let's

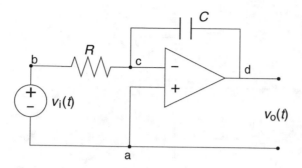

FIGURE 7.12: An integrator circuit.

calculate $V_o(t)$ in terms of $V_i(t)$. Writing the current law at the inverting terminal.

$$I_{bc} + I_{dc} = I_-$$

I_{bc} is the current through a resistor so we can write it in terms of the voltage across the resistor using Ohm's law. I_{dc} is the current through a capacitor so we use the capacitor component rule to write I_{dc} in terms of V_{dc}. The current I_- is zero.

$$\frac{V_{bc}}{R} + C\frac{d}{dt}V_{dc} = 0$$

Picking a to be the reference node and writing the voltages in terms of node voltages we get

$$\frac{V_{ba} - V_{ca}}{R} + C\frac{d}{dt}(V_{da} - V_{ca}) = 0 \qquad (7.6)$$

Using the virtual short assumption $V_{ca} = V_{aa} = 0$. Substituting $V_{ba} = V_i(t)$ and $V_{da} = V_o(t)$ into Eq. (7.6) we obtain

$$\frac{V_i(t)}{R} + C\frac{d}{dt}V_o(t) = 0$$

or

$$\frac{d}{dt}V_o(t) = -\frac{1}{RC}V_i(t)$$

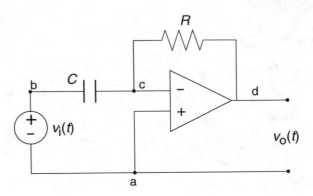

FIGURE 7.13: A differentiator circuit.

Integrating both sides

$$V_o(t) = -\frac{1}{RC}\int V_i(t)\,dt$$

Thus the output voltage is a scaled integral of the input voltage.

7.6.5 Differentiation

By switching the places of the resistor and capacitor as shown in Fig. 7.13 we can obtain a circuit whose output is the derivative of the input. Writing the current law at the inverting input we obtain

$$I_{bc} + I_{dc} = I_-$$

In this case I_{bc} is the current through the capacitor and I_{dc} is the current through the resistor.

$$C\frac{d}{dt}V_{bc} + \frac{V_{dc}}{R} = 0$$

Writing these voltages in terms of the node voltages with a being the reference node.

$$C\frac{d}{dt}(V_{ba} - V_{ca}) + \frac{V_{da} - V_{ca}}{R} = 0$$

Using the virtual short assumption $V_{ca} = V_{aa} = 0$. Substituting $V_{ba} = V_i(t)$ and $V_{da} = V_o(t)$ into the equation we get

$$C\frac{d}{dt}V_i(t) + \frac{V_o(t)}{R} = 0$$

or

$$V_o(t) = -RC\frac{d}{dt}V_i(t)$$

The integrator and differentiator circuits are used in many different application. One way they can be used is to simulate differential equations. As differential equations are used to model a whole variety of complex systems, including economic systems, chemical systems, mechanical systems and of course, electrical systems, this is a very useful application.

7.7 SUMMARY

In this chapter we have introduced the operational amplifier, a complex and useful component with a surprisingly simple component rule. We have shown how the op-amp can be used to amplify voltages, to act as a buffer, and to implement various algebraic operations.

7.8 PROBLEMS

1. In the following circuit find V_{out}.

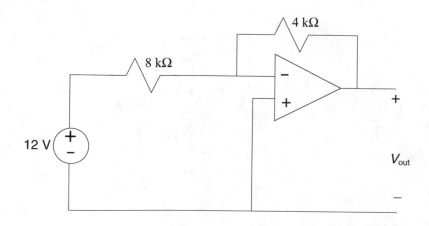

2. In the following circuit find V_{out}.

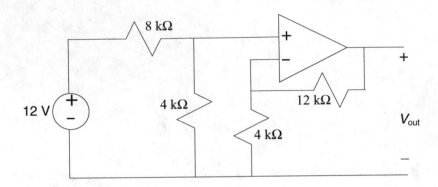

3. In the following circuit find V_o.

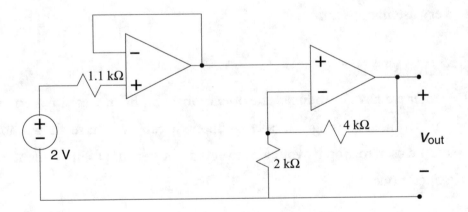

4. In the circuit below find V_{out}.

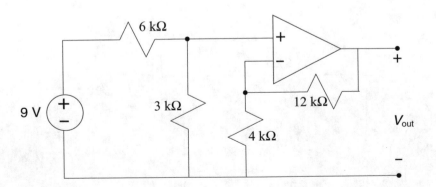

5. In the following circuit find V_{out}.

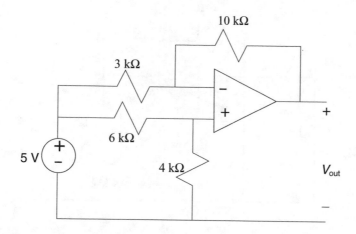

6. In the following circuit find V_{out}.

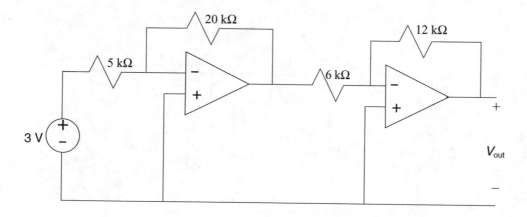

7. In the following circuit find V_{out}.

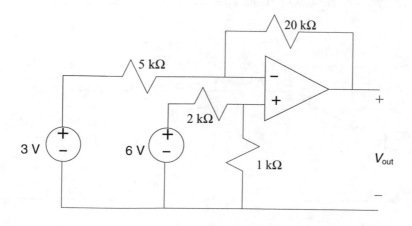

8. In the following circuit find V_{out} when $V_{in} = 0.5$ V.

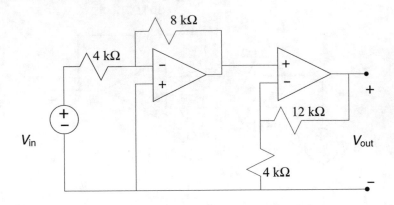

9. In the following circuit find V_o.

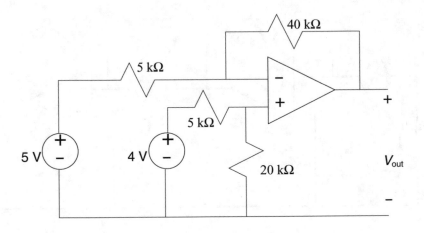

10. In the following circuit find V_{out} when $V_{in} = 0.5$ V.

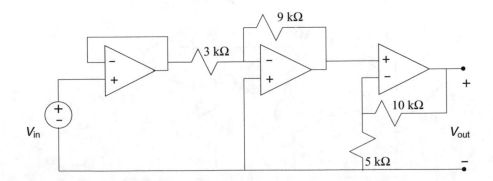

11.   In the circuit below find I_c and V_{ce} if $V_{in} = 0.5$ V.

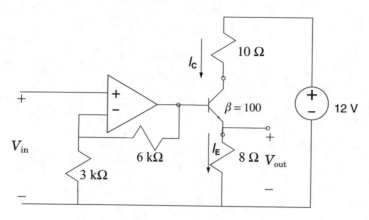

Printed in the United States
by Baker & Taylor Publisher Services